Modeling of Wind Turbines with Doubly Fed Generator System

Jens Fortmann

Modeling of Wind Turbines with Doubly Fed Generator System

 Springer Vieweg

Jens Fortmann
Duisburg, Germany

Dissertation University Duisburg-Essen, 2014

ISBN 978-3-658-06881-3   ISBN 978-3-658-06882-0 (eBook)
DOI 10.1007/978-3-658-06882-0

The Deutsche Nationalbibliothek lists this publication in the Deutsche Nationalbibliografie; detailed bibliographic data are available in the Internet at http://dnb.d-nb.de.

Library of Congress Control Number: 2014948095

Springer Vieweg
© Springer Fachmedien Wiesbaden 2015

Printed on acid-free paper

Springer Vieweg is a brand of Springer DE.
Springer DE is part of Springer Science+Business Media.
www.springer-vieweg.de

Acknowledgements

This thesis resulted from my part-time activities at the department for Electrical Power Systems, University of Duisburg-Essen. First, I want to thank Matthias Schubert, Peter Quell and Jörg Zeumer who accepted and supported the idea of an industrial PhD and were willing to support it even through some turbulent times.

My special and heartfelt thanks to Prof. Dr.-Ing. habil István Erlich, head of the department for Electrical Power Systems, University of Duisburg-Essen, for the thoughtful guidance and support of this work. His suggestions and comments, in many interesting technical discussions have contributed essentially to the success of this work.

Many thanks go also to Prof. i.r. W.L Kling of the Technische Univesiteit Eindhoven for agreeing to act as second assessor of this thesis.

Furthermore, I would like to thank my colleagues at Senvion SE, the staff of the department of Electrical Power Systems, University of Duisburg-Essen and my colleagues at Woodward for their support and many interesting discussions.

Finally and most important, I want to thank my wife Sabine and my son for their support and understanding over the last years.

Berlin   Jens Fortmann

Contents

Table of Figures

Nomenclature

Notation

Notation Type	Example	Explanation
Italic	I, l, x	Scalar physical quantity or numerical variable (i, j)
<u>underlined</u>	$\underline{V}, \underline{z}$	Phasors and complex values are represented using underline
CAPITAL	V, I	Capital letters denote values in physical (SI) units.
small letters	$i, r, \underline{z}$	Small letters denote normalized quantities, the same as instantaneous variables. The time dependency of instantaneous values is sometimes emphasized using the notation $v(t),\ i(t), \psi(t)$
bold small letter	**a, h**	Bold small letters denote a vector
BOLD CAPITAL	**A, H**	Bold capital letters denote matrices
Roman	Hz, A	Unit symbols are written in roman type
Roman	sin, tan	Standard mathematical functions are written in roman type
Roman	$\mathrm{d}f/\mathrm{d}t,$ $\partial p / \partial x$	Derivatives and partial derivatives are written in roman type
Roman	1,2,3	Numbers are written in roman type
Roman	v_R	Subscripts denoting objects are written using roman type

General Definitions

$\underline{V}, \underline{I}, \underline{\Psi}$	Complex RMS voltage, current, flux linkage RMS quantities are represented using capital letters and underlined when complex.
$v/\underline{v},\ i/\underline{i},\ \psi/\underline{\psi}$	Instantaneous scalar/complex voltage, current, flux linkage. Small letters are used for instantaneous values, and the time dependency is sometimes emphasized using the notation $v(t),\ i(t), \psi(t)$.
$\hat{v},\ \hat{i},\ \hat{\psi}$	Amplitude of the sinusoidal voltage, current, flux linkage

Symbols

A_WR	Swept rotor area
E_w	Kinetic energy of the wind
H	Inertia constant of the rotating masses of the WT
D_Shaft	Damping of the turbine shaft
J_WR	Rotational inertia of wind turbine rotor
J_R	Rotational inertia of wind generator rotor

K_{DT}	Stiffness of the turbine drive train assuming a stiff rotor
K_{Shaft}	Resulting modified stiffness of the turbine drive train
F_L, F_D	Lift and drag force of rotor blade
P	Power
P_W	Aerodynamic power
R_{WR}	Wind turbine rotor radius
Θ	Pitch angle
α_A	Aerodynamic angle of attack
c_L, c_D	Lift and drag coefficients of rotor blade
c_P	Aerodynamic power coefficient
f_{Edge}	First edgewise eigenfrequency of the turbine rotor blades
$\underline{i}_S, \underline{i}_R$	Complex stator, rotor currents
$l_{\sigma S}, l_{\sigma R}$	Stator, rotor leakage inductances; as p.u. quantities, they correspond with the reactances $x_{\sigma S}, x_{\sigma R}$
l_m	Mutual (main)-field inductance; as p.u. quantity; it corresponds with the reactance x_m
m	Mass
λ	Tip speed ratio
λ_n, λ_{opt}	Tip speed ratio at nominal speed, tip speed ration at optimal c_P value
ω_R, ω_0	Rotor angular speed, synchronous speed
φ	Electrical angle
ρ	Air density
$\underline{\psi}_S, \underline{\psi}_R$	Complex stator, rotor flux-linkages
r_S, r_R	Stator and rotor resistances
s	Slip of induction machines (small letter); in a transfer function "s" means Laplace operator
t_{el}, t_m	Electrical/mechanical torque
$\underline{v}_S, \underline{v}_R$	Complex stator, rotor terminal voltages
$\underline{v}_{DC}$	Converter DC-link voltage
v_w	Wind speed
v_{wi}	Induced wind speed
v_{wr}	Wind speed component caused by wind turbine rotor speed
v_{w1}, v_{w2}, v_{w3}	Wind speed before, in and behind the rotor plane
$\underline{z}' = (r_S + j\omega_0 l')$	Transient impedance of the induction machine

Superscripts/Subscripts

a, b, c	Original components of a three phase system
α, β	Direct, quadrature axis component in stationary reference frame
d, q	Direct, quadrature axis component in rotating reference frame

1, 2, 0	Positive, negative and zero sequent components
P, Q	Active and reactive component of current (capital letters)
S, R, WR	Stator, (generator) rotor, wind (turbine) rotor
m, σ	Mutual field, leakage
$\angle$	Used to characterize the reference frame; without extension indicates arbitrary reference frame
$\angle 0$	Stationary reference frame, e.g. $i_{Sd}^{\angle 0}$, $i_{Sq}^{\angle 0}$, $\underline{i}_{S}^{\angle 0}$
	Note: $i_{Sd}^{\angle 0} = i_{S\alpha}$, $i_{Sq}^{\angle 0} = i_{S\beta}$
$\angle \omega_0$	Synchronously rotating reference frame, e.g. $i_{Sd}^{\angle \omega_0}$ $i_{Sq}^{\angle \omega_0}$, $\underline{i}_{S}^{\angle \omega_0}$
$\angle v_S$	Reference frame where the d-axis is oriented along the stator voltage, e.g. $i_{Sq}^{\angle v_S}$, $\underline{i}_{S}^{\angle v_S}$
$\angle \psi_S$	Reference frame where the d-axis is oriented along the stator flux, e.g. $i_{Sq}^{\angle \psi_S}$, $\underline{i}_{S}^{\angle \psi_S}$

Abbreviations used

DFG	Doubly Fed Generator system
FSC	Full Size Converter system
MSC	Machine Side Converter
LSC	Line Side Converter, grid side converter
OLTC	On-Load Tap Changer
PCC	Point of Common Coupling
WP	Wind Plant, wind power plant
WT	Wind Turbine, wind turbine terminals

1 Introduction

1.1 Motivation

Wind energy is playing an increasingly important role in the supply of energy of most industrialized countries, and its share of the electrical generation is expected to continue to increase for many years to come. The EUs target for 2020 is a 20% share of energy from renewables; it is assumed that a high percentage of the expected growth will be from wind power [1]. Recent findings suggest that the share of 20% may even be exceeded at least in some regions [2], see Fig. 1.1.

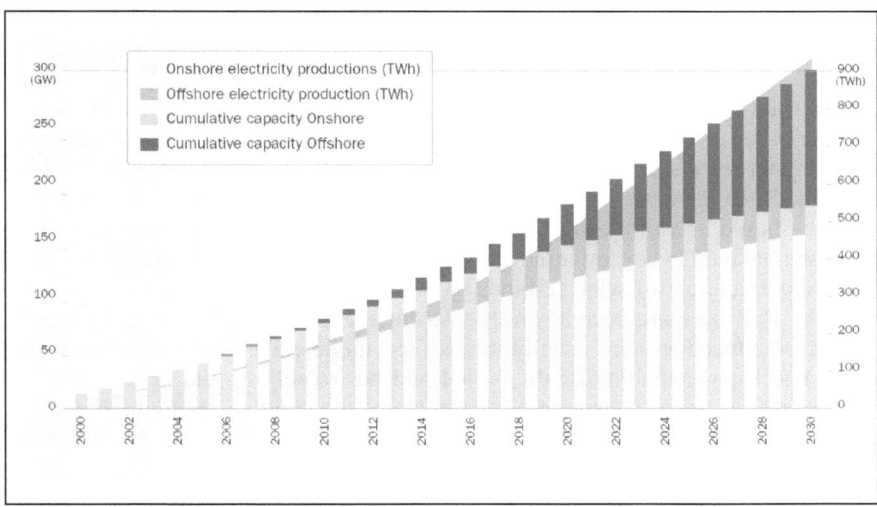

Fig. 1.1 Projected growth of wind electricity generation (source: EWEA, Pure Power Scenario).

From a grid stability point of view, not the energy but the power is the critical figure. During low load, high wind conditions, the share of the wind energy production could reach 100% demand in several regions [3], see Fig. 1.2. Existing power stations do not only provide power, there is a range of so-called ancillary services that are also provided like frequency control, voltage control, reserve power provision and the capability to temporarily operate outside the normal operating conditions following events in the grid.

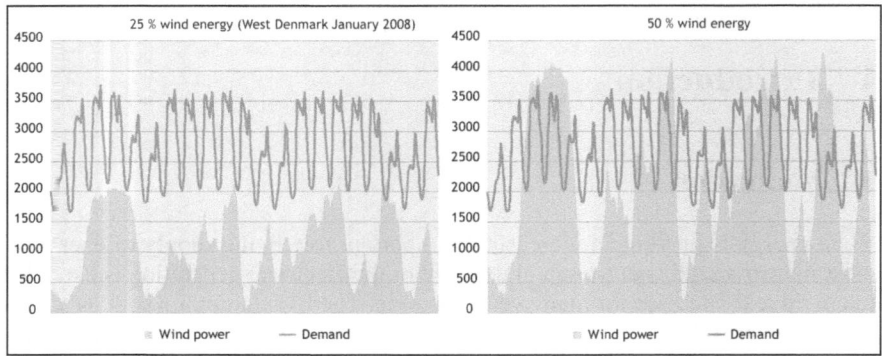

Fig. 1.2 Wind power production and demand, measurement and assumption for growth to 50%. Source: EcoGrid [3].

One possibility would be to keep conventional power stations in operation to provide e.g. the necessary reactive power for a stable operation in the case of grid faults. But this will probably not make sense from an economic point of view.

In order to analyze system stability and to develop new operation strategies with an increased share of renewable energy for the grid, there is a need for reliable models to simulate the response of wind turbines to events in the grid. The models should be valid to study both short term events like grid faults and conditions that take up to several minutes like the analysis of a voltage collapse.

Manufacturers of wind turbines commonly have very detailed wind turbine models. But there are several drawbacks for the use of such models by system operators. Since these are usually 'black-box' models, the grid operator does not have a full understanding of the model and has to rely on the manufacturer in case questions arise. A change or an update of the simulation environment may even render the model unusable. A model update could then only be performed by the manufacturer, but the result may not be available within the time scale desired by the grid operator.

As a result, so called generic models have been required by grid operators in several countries. The idea behind these models is that the structure of the model is fully documented, and that model implementations are available in different simulation environments. Once the model is developed, a manufacturer only needs to provide a parameterization for the model. A drawback of generic models can be the possibly reduced accuracy compared to manufacturer specific models. A difficult task for generic model development is therefore to find the right balance between desired accuracy and model complexity.

The first steps of model development are (a) to describe the physical background of the turbine components and (b) to analyze the relationship between the level of modeling detail and the impact on the active and − if applicable − reactive current output of the turbine. As it can be shown, an increase in model complexity will in many cases lead only to very small improvements in modeling accuracy.

It is therefore a key aim of this work to describe, which parts of a model description are actually required and which parts do not provide a relevant contribution to model accuracy and can be omitted.

1.2 Wind Power Conversion Systems

There are different wind turbine concepts available today. One of the key differences is the way, the active power is controlled. The simplest way used mainly in turbines with fixed speed is the passive *stall control*. The rotor blades are mounted at a fixed blade pitch angle. If the wind speed increases beyond a defined point, a reduction of blade efficiency through aerodynamic stall reduces the aerodynamic power. Stall control is common for small wind turbines, but higher loads and limited control capabilities compared to other strategies usually limit the application to wind turbines below 1.3 MW .. 1.5 MW.

The most common way to control the aerodynamic power for large wind turbines is by turning the blades to feather position once rated wind speed is reached. This is referred to as *pitch control*. It is the most common approach today. In contrast to stall control, it allows an easy implementation of variable speed operation.

A modification of stall control is called *active stall control*. Like pitch control the blades can be turned, but unlike pitch the movement of the blades is away from feather position to improve the effect of stall at above rated wind speed. If used with fixed speed turbines, active stall control leads to smaller fluctuations in aerodynamic power than pitch control. But it tends to lead to higher loads compared to variable speed pitch controlled wind turbines and is therefore common only for smaller, fixed speed turbines.

1.2.1   Wind Turbine Classification

A classification of the most relevant wind turbine types based on their electrical connection the grid is shown in Fig. 1.3. Even though some more combinations exist (for example fixed speed pitch or variable speed active stall control), they do not play a role in practice any more.

Type 1 turbines operate at fixed speed using squirrel cage induction generators. The aerodynamic power control is achieved using passive aerodynamic stall or active stall blade control. Capacitor banks or additional active power electronics are used to control reactive power.

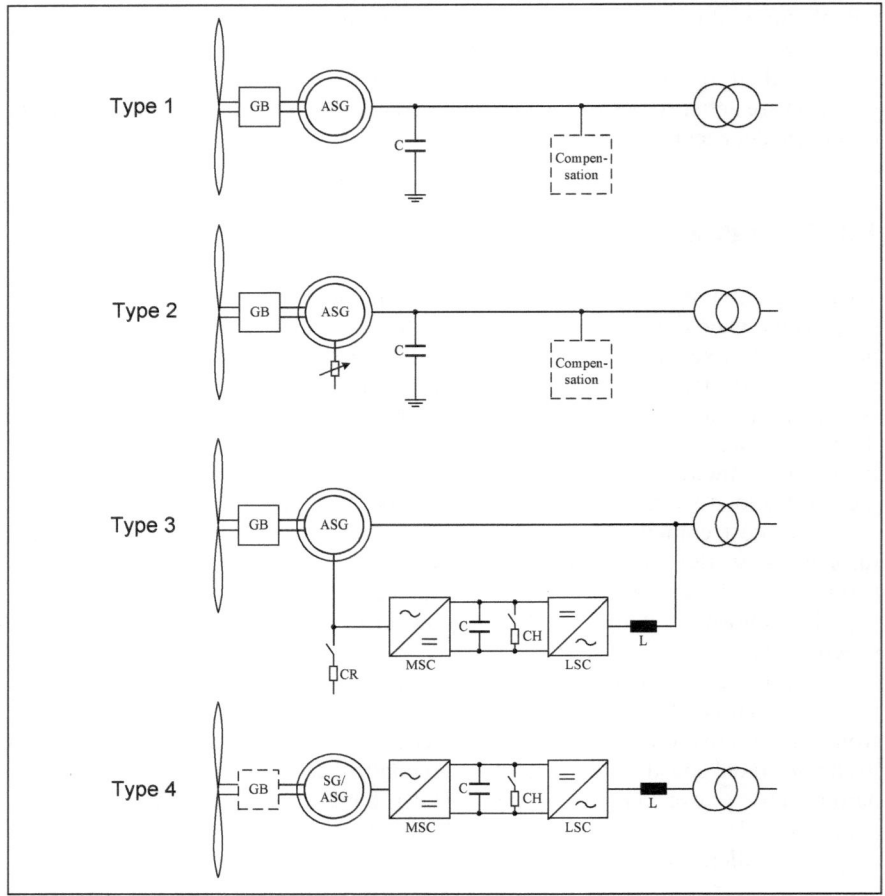

Fig. 1.3   Classification of wind turbines based on their electrical connection to the grid.

Type 2 turbines use a wound rotor induction generator and operate within a limited operating range. The variation of the speed is achieved using controlled resistors in the rotor circuit of the generator. Pitch control is generally used for aerodynamic power limitation. Capacitor banks or additional active power electronics are used to control reactive power.

Type 3 turbines use a wound rotor induction generator and a converter on the rotor side that is rated for partial load. This allows an independent control of active and reactive power within a wide speed operating range. Pitch control is used for aerodynamic power limitation. This turbine type is referred to as doubly fed generator system (DFG) since both rotor and stator can feed power into the grid[1].

Type 4 turbines are based on converters connected to the stator side of the generator that are rated at full turbine power. They are referred to as full size converter (FSC) turbines. Pitch control is generally used for aerodynamic power limitation. Different generator types are common, wound rotor synchronous generators, permanent magnet synchronous generators and squirrel cage induction generators. Depending on the generator, a gearbox may or may not be used.

1.2.2 Wind Turbine Model Structure

The structure of wind turbine models depends very much on the focus of the models. Very detailed aerodynamic models and corresponding wind models are necessary for wind turbine design and for the mechanical type certification. For the use in electrical dynamic simulation studies, such a detailed aerodynamic model is usually not necessary.

An overview of the model structure used for describing the wind turbine is given in Fig. 1.4. It is based on the structure presented in [4]. Using this structure, the level of detail of the individual blocks is still open. This allows a very modular approach to simulation. If required, individual blocks can be replaced by more detailed blocks in order to study specific effects or situations.

The *wind turbine aerodynamic and drive train* block contains both the entire mechanical representation of the wind turbine including the pitch drive and the aerodynamic model that describes the conversion of wind speed to mechanical power. The mechanical model can embody either a simple single mass representation of the wind turbine drive train or a higher order mechanical representation of the wind turbine. Inputs to this block are the wind speed, the pitch angle demand and the mechanical power from the generator. Output of the block is the generator speed.

The *generator and converter* block represents the electrical model of the generator and the converter including machine side converter and grid side converter and, if available, DC-link energy absorbers. Inputs to the generator model are the generator speed, active and reactive current reference and the grid voltage. Output of the generator model is the active and reactive power that is fed into the grid and the mechanical shaft power of the generator.

---

[1] Other common names are doubly fed induction generator (DIFG), doubly fed asynchronous generator (DFAG) and doubly fed induction machine (DFIM).

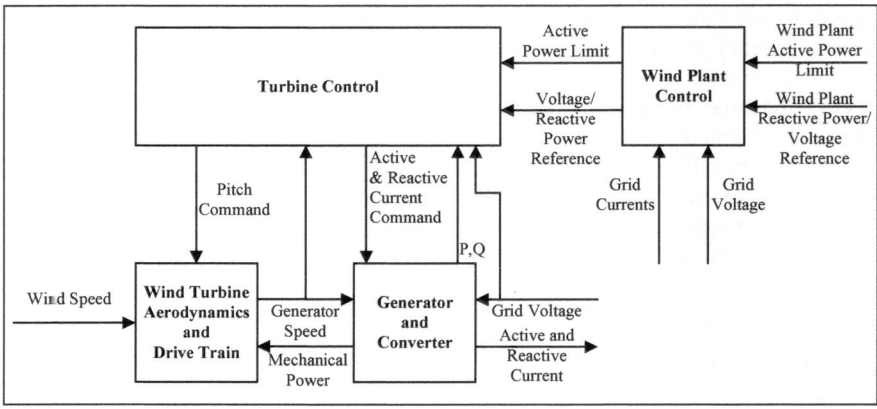

Fig. 1.4   Wind turbine model structure.

The *turbine control* block consists of the wind turbine controller that is responsible for controlling rotor speed and active and reactive power of the turbine. Inputs to this block are generator speed, turbine voltage and active and reactive power output of the turbine on one side and reference values from the wind plant control on the other side. Outputs of the turbine controller are the reference angle for the pitch system and reference values for active and reactive currents of the converter. In practical implementations, functions like drive train damping and current reference calculation during grid faults could be located either in the converter or in the turbine control. For clarity of the model structure, these functions are concentrated in the control block.

The *wind plant control* block contains the wind plant controller that receives reference values from the grid operator and sends reference values to the individual wind turbines. A wind plant controller usually consists of an active power control loop and a reactive power control loop. Only the reactive power control of the plant controller is considered in this work; any active power curtailment demand by external reference from a grid operator is forwarded directly to the turbines without an additional control loop at plant level.

1.3 Aim of the Work

This thesis is focused on two main fields,

(1) the development of a new generic model of a wind turbine for the use in grid simulation studies that is based on a physical deduction of the models of all components, and

(2) the design of a control system for wind turbines and wind plants that has improved control capabilities with respect to voltage/reactive power compared to conventional power stations.

1. Generic Model Development

One approach to model a system is to represent as much details as technically possible in a mathematical form. Such a kind of model can lead to rather detailed results. The drawbacks are high computational requirements, slow model execution and the need for a detailed knowledge to provide the parameters required.

Existing generic models by contrast try to use a very simple representation that is not based on a direct physical background but instead uses empirical approaches that correlate the model to measurements or more detailed models.

Key intentions of the generic model development in this work are

(a) to identify the minimum level of detail that is required for dynamic modeling of wind turbines

(b) to base all simplifications of a model on the turbine physics, that means if possible all parameters should have a physical equivalent or should be deductible from physical quantities or control parameters in the case of the turbine control.

(c) to develop a generic model of a wind turbine that is independent of a manufacturer specific implementation, with focus on creating new generic sub-models for the aerodynamic system, the control system and the generator/converter system.

(d) to develop a model that can be used both for DFG (type 3) and FSC (type 4) wind turbines.

2. Wind Plant Reactive Power Control Design

Wind turbines in wind plants connected to the high voltage system replace synchronous generators. In future, it will be necessary to operate grids with limited or even no synchronous generation. Newer grid codes have started to become more explicit in demanding some of the implicit capabilities of synchronous generators explicitly, for example the (inherent) capability to stabilize the voltage by providing reactive current in response to fast voltage changes. But even newer grid codes (implicitly) assume the existence of a sufficient base capacity of synchronous generation to ensure a stable operation of the grid.

Converter based wind turbines with DFG or FSC have a far superior control capability compared to synchronous generators. Based on the analysis of the dynamic response of synchronous generators and an analysis of the grid requirements, a hierarchical control structure for wind plants and wind turbines has been developed that can provide an improved performance with respect to voltage and reactive power compared to conventional power stations based on synchronous generators.

Requirements for the response to grid faults and voltage step changes of wind turbines are usually specified and implemented at turbine level. Due to the wind plant collector system, possibly different high voltage transformers and the distributed nature of wind turbines, a comparison of a response of synchronous generators and wind plants at medium voltage level is not sufficient. The high voltage terminals are therefore seen as the reference point in this work.

Key intentions of the reactive power control design in this work are

(a)   to define dynamic reactive current requirements at the high voltage terminals based on an analysis of the grid requirements

(b)   to design a hierarchical wind plant and wind turbine control system that can provide an improved dynamic reactive current response at the high voltage terminals compared to the existing power system

(c)   to show the capability of the proposed controller design for electrically close and distant faults and for voltage step changes compared to existing power stations based on synchronous generators

(d)   to compare the response of the proposed control structure to existing implementations of wind turbine and wind plant reactive power control.

1.4 Structure of the Work

Chapter 1 gives a general overview of the work and shows the motivation.

Chapter 2 describes the aerodynamic models for the dynamic simulation of wind turbines. A generic aerodynamic model is presented that allows the replacement of commonly used c_P-λ-Θ tables in dynamic simulation models. The approach is compared to other existing concepts and can reduce the complexity of the aerodynamic representation.

Chapter 3 analyzes the degree of detail necessary to model of the key mechanical components of a wind turbine. The possible gain of accuracy for dynamic models using higher order mechanical representations is compared to errors resulting from the use of single mass and two mass representations of the drive train.

Chapter 4 describes the key concepts necessary to model the primary turbine control. Different approaches for blade pitch angle and generator torque control are compared and a generic approach is presented.

Chapter 5 describes generator and converter models for DFG and FSC based wind turbines. The model is derived from the physical equations of the generator, the converter hardware and converter control system. A generic model is derived that is more accurate than existing generic models but maintains a simple structure. The model is set up in a way that in can be used both for DFG and FSC type wind turbines.

Chapter 6 describes the design of a reactive power voltage control for wind plants. Based on an analysis of the grid requirements at the high voltage terminals, a hierarchical control structure and two alternative control implementations for wind turbines are derived. The response of the proposed control structure to grid faults and voltage step changes is compared to synchronous generators and wind plants using other control approaches.

Chapter 7 summarizes the contents of the work and gives an outlook for the application of the proposed model and reactive power/voltage control structure.

2 Model of the Turbine Aerodynamics

2.1 Introduction

The primary time domain of interest for dynamic grid studies is up to 10 s .. 30 s, until the key disturbances have either settled or lead to a disconnection of parts of the system. Within the time frame given, wind speed is often assumed to be constant [4]. Short term changes of wind speed in the range of seconds like gusts and turbulent wind generally do not have an impact on stability for variable speed wind turbines as they have local effects only that average out even within one wind plant.

Using a single, aggregated model of the wind power plant instead of a detailed representation of each wind turbine is usually acceptable for stability studies [5], [6]. However, the analysis of large blackouts and system disturbances like 2003 in Italy [7] or 2006 in the UCTE System [8] on the other hand shows that the time frame of interest could be up to several minutes. In these cases an aerodynamic model is required that allows simulation of the excursion of aerodynamic quantities like blade pitch angle, wind speed and wind turbine rotor speed in a simple form.

More detailed aerodynamic models are used for the modeling of fixed speed turbines [9], [10]. An overview of the aerodynamic model within the turbine model structure is shown in Fig. 1.4.

2.2 Energy Capture from the Wind

The theory of extracting energy from wind dates back to Betz [11]. The kinetic energy of wind E_W and the resulting power P_W are defined by

$$E_\mathrm{w} = \frac{1}{2} m_\mathrm{w} v_\mathrm{w}^2 \qquad (2.1)$$

$$P_\mathrm{w} = \dot{E}_\mathrm{w} = \frac{1}{2} \dot{m}_\mathrm{w} v_\mathrm{w}^2 = \frac{1}{2} \rho A v_\mathrm{w}^3 \qquad (2.2)$$

with the air flow $\dot{m}_\mathrm{w} = \rho A_\mathrm{WR} v_\mathrm{w}$, ρ as air density and A as area through which the air flows. Assuming a homogenous tube, with wind speed v_{W1} at the entry and wind speed v_{W3} at the exit, and a wind speed v_{W2} in between (see Fig. 2.1), the following equation

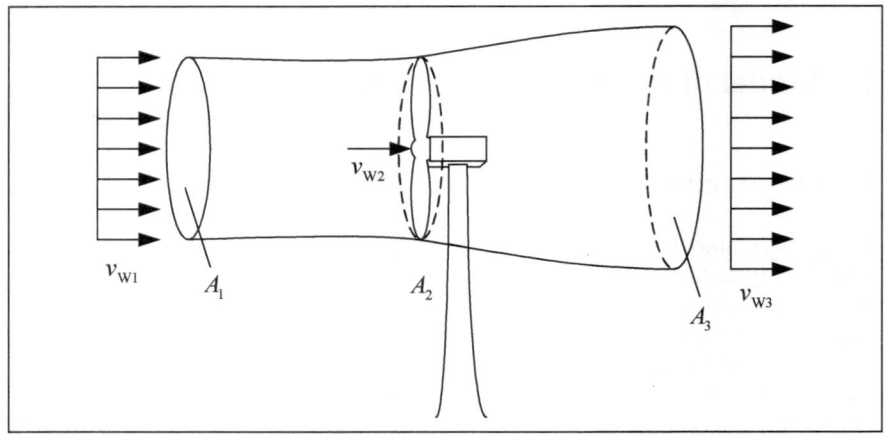

Fig. 2.1   Air flow at a wind turbine.

$$\rho A_1 v_{W1}^2 = \rho A_2 v_{W2}^2 = \rho A_3 v_{W3}^2 \tag{2.3}$$

is true if the air density ρ is assumed to be constant before and after the tube. The energy and the power that can be extracted from the wind then are

$$E_W = \frac{1}{2} m_W (v_{W1}^2 - v_{W3}^2) \tag{2.4}$$

$$P_W = \dot{E}_W = \frac{1}{2} \dot{m}_W (v_{W1}^2 - v_{W3}^2) \tag{2.5}$$

with the flow in the rotor plane as

$$\dot{m}_W = \rho A_2 v_{W2} \tag{2.6}$$

Applying the Froud-Rankine theorem ([12], S.184) it can be proven, that the wind speed in the rotor plane is the average of the wind speed in front and behind the rotor plane, that means

$$v_{W2} = \frac{v_{W1} + v_{W3}}{2} \tag{2.7}$$

By inserting (2.6) with v_{W2} according to (2.7) into (2.5) the energy extracted by the wind for a given wind turbine rotor plane $A_{WR} = A_2$ (see Fig. 2.5) can then be described as

$$P_{\text{WR}} = \underbrace{\frac{1}{2}\rho\pi R_{\text{WR}}^2 v_{\text{W1}}^3}_{\text{wind power}} \underbrace{\left[\frac{1}{2}\left(1+\frac{v_{\text{W3}}}{v_{\text{W1}}}\right)\left(1-\left(\frac{v_{\text{W3}}}{v_{\text{W1}}}\right)^2\right)\right]}_{\text{power coefficient } c_p} \tag{2.8}$$

with R_{WR} as rotor radius. By substituting $x = v_{\text{W3}}/v_{\text{W1}}$ and setting the first derivative with respect to x of the resulting function

$$c_p(x) = \frac{1}{2}(1+x)(1-x^2) \tag{2.9}$$

to zero, we find a maximum for $x = 1/3$. From (2.7) we see that the maximum power extraction can be achieved by a wind speed in the rotor plane of $v_{\text{W2}}=2/3\,v_{\text{W1}}$. By inserting this result into (2.8) the maximum power coefficient c_P according to Betz is calculated as

$$c_{P,\text{Betz}} = \frac{16}{27} = 0.593 \tag{2.10}$$

This constant describes the maximum energy that can be extracted from the wind. The power factor describes a theoretical limit that is independent of factors like number of blades or blade design. With modern blades, a c_P of about 0.5 can be achieved at the optimum operation point.

2.3 Aerodynamics of Rotor Blades

For calculating the aerodynamic behavior of a turbine rotor, the blades are divided into different segments Δr. For each segment, lift and drag forces F_L and F_D can be described as functions of the angle of attack α_A between the induced wind speed v_{Wi}, and the blade chord. The induced wind speed v_{Wi} results from the reduced wind speed in the rotor plane $v_{\text{W2}} =2/3\,v_{\text{W1}}$ and the wind component caused by the rotor speed $v_{\text{Wr}} = \Omega_{\text{WR}} \cdot R_j$ with R_j as the distance from the hub center to the segment with the index j. For an easier description, the index is omitted in the following text. In Fig. 2.2 the angle of attack α_A is shown as a result of wind speed v_{W1} in front of the rotor, the wind speed induced by the rotor speed v_{Wr} and the angle Θ of the blade chord with respect to the rotor plane. Differential lift dF_L and drag dF_D forces and can be calculated as

$$dF_L = \frac{\rho}{2}v_{\text{Wi}}^2 c \cdot dr \cdot c_L(\alpha_A) \tag{2.11}$$

$$dF_D = \frac{\rho}{2} v_{Wi}^2 \cdot c \cdot dr \cdot c_D (\alpha_A) \tag{2.12}$$

with dr as the length of the profile in radial direction and c as the depth of the blade segment. Lift and drag coefficients c_L and c_D depend on the profiles selected and can be described as functions of the angle of attack α_A.

The induced wind speed v_{Wi} increases along the rotor from the hub to the blade tip. As a result, in order to achieve a constant angle of attack, the blade segments are twisted along the rotor axis. The profiles selected also change with increasing induced wind speed along the blade. The resulting in-plane force for each segment is the basis for the calculation of the captured energy, it can be written as

$$dF = \frac{\rho}{2} v_{Wi}^2 \cdot c \cdot dr \left[c_L (\alpha_A) \sin \alpha - c_D (\alpha_A) \cos \alpha \right] \tag{2.13}$$

with α as the angle between induced wind speed and the rotor plane (see Fig. 2.2).

For detailed simulations, lift and drag as function of the angle of attack are calculated for each profile used within the rotor blade. In addition, losses resulting from the spin induced into the wind by the rotor blades [13], losses due to the blade tips [12] and losses resulting from the chosen blade profiles have to be included in the calculation. A more detailed analysis of aerodynamic theory and blade design can be found in [12], [14] and [15]. Using detailed models, the input data for simplified representations as shown in the next section can be calculated.

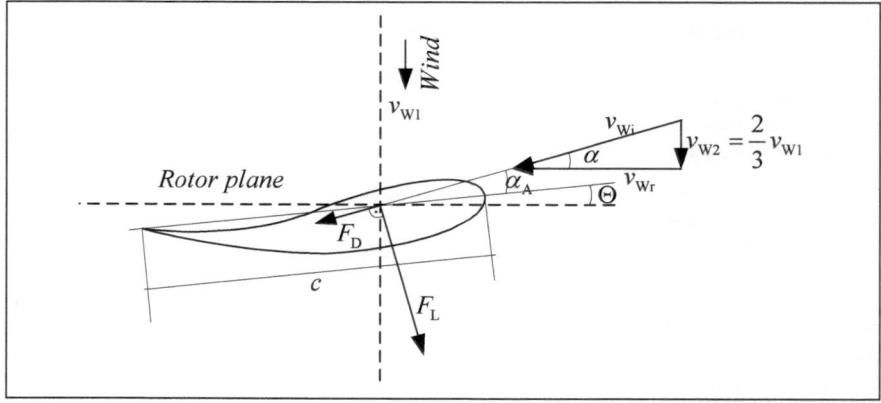

Fig. 2.2   Lift and drag of a blade in the rotor plane.

2.3.1 Simplified Representation

The energy captured by a rotor is calculated as the sum of the energy capture of each profile segment. Under steady state operating conditions, the definition of the dimensionless tip-speed ratio

$$\lambda = \frac{\Omega_{\mathrm{WR}} R_{\mathrm{WR}}}{v_{\mathrm{W}}} \tag{2.14}$$

with R_{WR} as rotor radius and Ω_{WR} as rotor speed allows a simplified representation of the resulting power coefficient of a wind turbine rotor as function of blade angle Θ and tip-speed ratio λ. The power generated by the wind turbine rotor is then calculated as

$$P_{\mathrm{WR}} = \frac{1}{2} \rho \pi R_{\mathrm{WR}}^2 v_{\mathrm{W}}^3 \cdot c_{\mathrm{P}} \left(\lambda, \Theta \right) \tag{2.15}$$

where $c_{\mathrm{P}} \left(\lambda, \Theta \right)$ is usually given as a table calculated by detailed aerodynamic simulation based on the blade element theory for a given blade design.

Consider, that the wind speed v_{W} corresponds with v_{W1} used in (2.3) - (2.8). In Fig. 2.3 the power coefficient is shown as a function of tip-speed-ratio λ for different blade angles. It can be seen that there is an optimum tip-speed-ratio λ_{opt} which results in a maximum power generation for a given wind speed at about 9 for a blade angle of 0 deg. The value λ_{n} corresponds to the operation at rated wind speed.

For practical reasons, there is an upper limit of the tip speed in order to limit noise and mechanical stress. There is also often a lower speed limit due to a limited operating range of the electrical system and in order to avoid resonances between the tower eigenfrequency and the blades. An ideal steady state operation trajectory that guarantees quasi optimal exploitation for wind is shown as black curve in Fig. 2.3. λ_{max} is the operating point at the lower rotor speed limit and the lowest wind speed used for wind turbine operation.

The same ideal steady state operation trajectory is shown in Fig. 2.4 for different wind speeds. Below rated wind speed (12 m/s in this case) the blade angle remains at zero. At higher wind speeds the blade angle is increased and thus the power is limited to the turbine rating. An operation with a blade angle of zero at higher wind speeds would allow the extraction of more power from the wind, but at the expense of a higher rating of the electrical system and the need for a stronger mechanical design. But such high wind speeds may only occur very seldom throughout a year. For economic reasons, the rated wind speed of a turbine is therefore always limited.

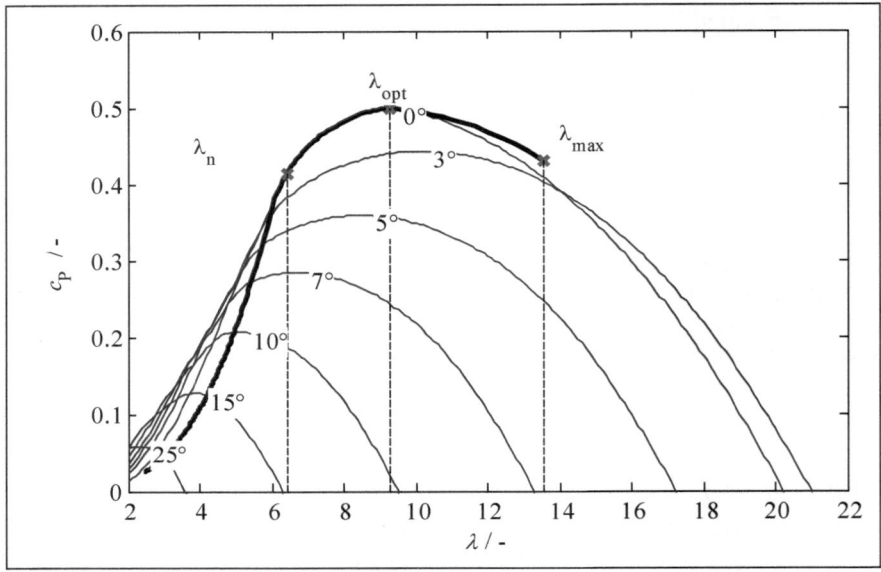

Fig. 2.3   Power coefficient c_P for different blade angles (thin) and steady state operation
trajectory (thick) of the wind turbine.

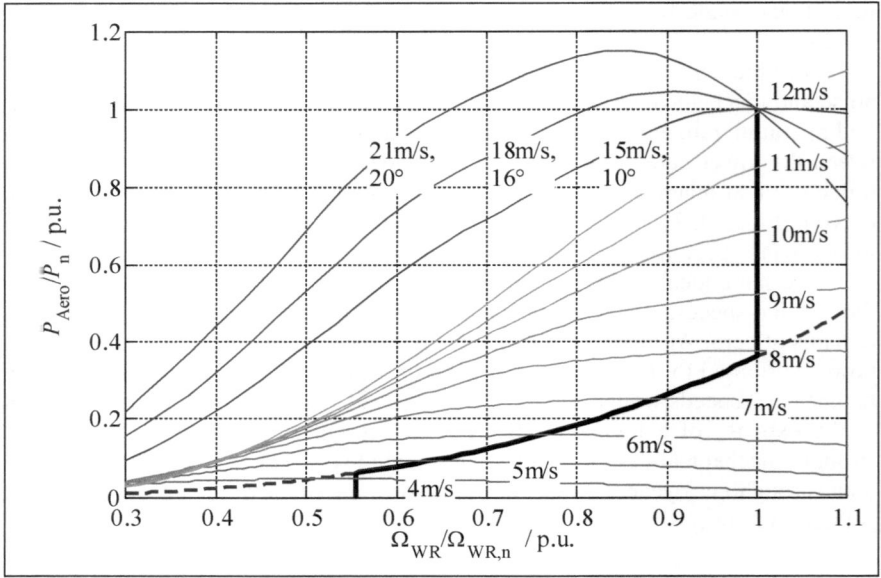

Fig 2.4   Wind turbine rotor power as a function of rotor speed for different wind speeds
(thin), optimal trajectory (dashed) and steady state operation trajectory (thick).

2.4 Simulation using c_P-λ Tables

The power coefficient c_P can be represented as a function of tip-speed-ratio λ and blade angle Θ as shown in Fig. 2.5. The thick line shows the operating trajectory for different wind speeds, starting from low wind speeds at high tip-speed-ratios to high wind speeds at low tip-speed-ratios. This is the same trajectory that can be seen in a different representation in Fig. 2.3 and Fig. 2.4. At very high wind speeds the power is limited to rated power by increasing the blade angle. At very low wind speeds and if the lower rotor speed is limited, an increase of the blade angle can increase the c_P –value and thereby the power captured from the wind to a certain degree. Describing c_P as function of λ and Θ (this is commonly referred to as c_P-λ table) allows calculating the power of the turbine in a wide operation range.

However, the c_P-λ table is usually large and therefore the method is not very efficient in simulations especially if many different wind turbines have to be represented. Besides, searching the right operating point in a two-dimensional table requires an interpolation in every simulation time step, this can also be time consuming. Therefore, a simplified representation of c_P is favorable if the accuracy requirements are not very challenging as it is usually the case in grid studies.

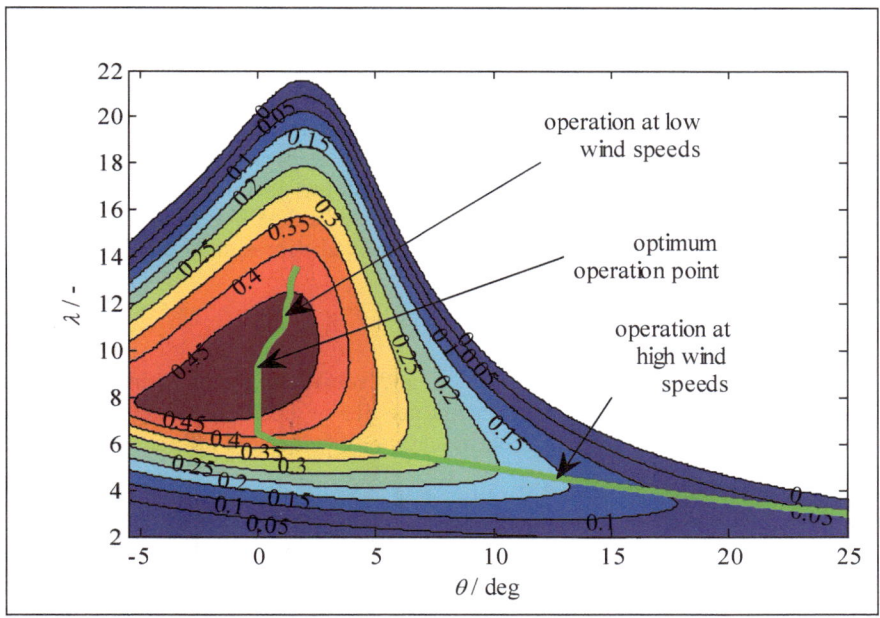

Fig. 2.5   Power coefficient c_P as function of blade angle Θ and tip-speed ratio λ.

2.4. Functional Representations for c_P-λ Tables

Some efforts have been published to reduce the number of parameters needed to represent the content of c_P-λ tables. A functional representation and a modification of this approach are presented here and analyzed later in section 2.5.9.

Functional Representation by Anderson and Bose

An often cited functional representation first published by [16] in 1983 and modified for the development of the GROWIAN by [17] in the form presented by [18] is

$$c_P(\Theta, \lambda) = c_1 \left(\frac{c_2}{\lambda_1} - c_3\Theta - c_4\Theta^{C_5} - c_6 \right) e^{c_7/\lambda_1} \qquad (2.16)$$

with

$$\frac{1}{\lambda_1} = \frac{1}{\lambda + c_8\Theta} - \frac{c_9}{\Theta^3 + 1} \qquad (2.17)$$

and the coefficients

$$c_{1-9} = \begin{bmatrix} 0.5 & 116 & 0.4 & 0 & 2 & 5 & 21 & 0.08 & 0.035 \end{bmatrix} \qquad (2.18)$$

Modified Functional Representation

A comparison with blade data used in modern turbines shows that the accuracy of (2.16) is not acceptable for modern turbines. The parameter sets given by the authors does not fit with modern blades good enough (see Fig. 2.6). A comparison of the representation with real turbine data of a 2 MW turbine shows that a rescaling of the blade angle allows an improved representation. This can be achieved with

$$\Theta' = c_{10}\Theta \qquad (2.19)$$

by adding and additional coefficient c_{10} and substituting Θ by Θ' in (2.16) and (2.17). In a second step, the coefficients need to be updated. Using a nonlinear parameter fit [19], the following coefficients were calculated for a 2 MW turbine:

$$c_{1-10} = \begin{bmatrix} 0.297 & 118 & -0.50 & 0.922 & 1.12 & 3.33 & 15.6 & 0.102 & 0.017 & 0.751 \end{bmatrix} \quad (2.20)$$

This modification leads to a significantly improved correlation of the c_P-λ data generated by (2.16) with real turbines (see Fig. 2.7) The results of these functions are further analyzed in section 2.5.9.

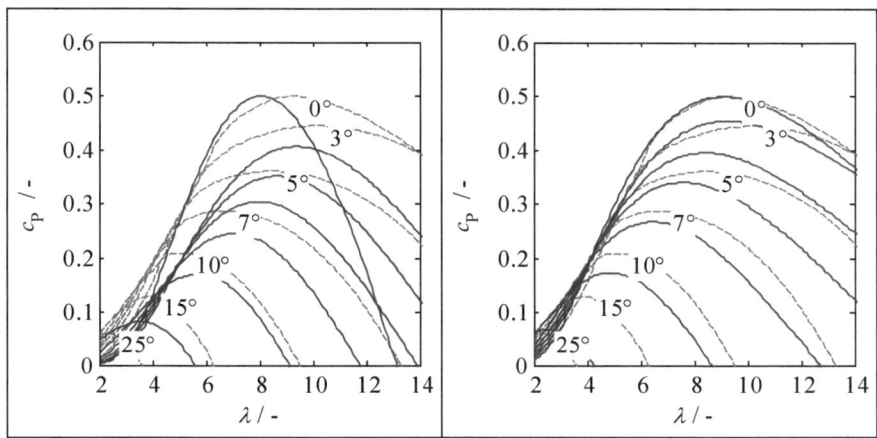

Fig. 2.6 c_P values for the original functional representation according to (2.16) (solid) and reference (dashed).

Fig. 2.7 c_P values for the modified functional representation according to (2.17) (solid) and reference (dashed).

2.4.2 Polynomial Fit Representation for c_P-λ Tables

A different approach for reducing the number of parameters compared to a c_P-λ table consists in using a polynomial fit representation [20]

$$c_P(\Theta, \lambda) = \sum_{i=1}^{n} \sum_{j=1}^{m} \alpha_{i,j} \Theta^i \lambda^j \qquad (2.21)$$

with the coefficients $\alpha_{i,j}$ and the order of the polynomial given by n and m.

It can be shown that an even number parameters leads to an improved representation at the outer limits of the function. A good representation within a limited range from $\lambda = [2; 13]$ and $\Theta = [0; 25]$ deg can be reached with a 4x4 representation using 16 coefficients.

$$
\left[\alpha_{i,j}\right] =
\begin{bmatrix}
8.4018e\text{-}008 & -8.2359e\text{-}007 & 9.8303e\text{-}006 & -1.8864e\text{-}005 \\
-8.4245e\text{-}006 & 7.5152e\text{-}005 & -5.8513e\text{-}004 & 7.3378e\text{-}004 \\
7.2467e\text{-}005 & -1.0032e\text{-}003 & 1.5504e\text{-}003 & 1.1218e\text{-}002 \\
5.8657e\text{-}005 & -9.0477e\text{-}003 & 1.6776e\text{-}001 & -3.3920e\text{-}001
\end{bmatrix}
\qquad (2.22)
$$

A comparison of the polynomial representation with the c_P-λ curves of a 2 MW turbine is shown in Fig. 2.8. The results of this function are further analyzed in section 2.5.9.

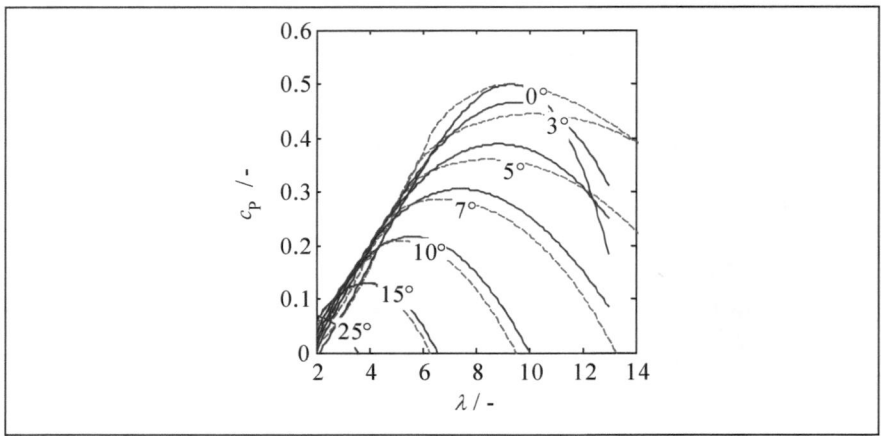

Fig. 2.8 c_P values for the 3rd order polynomial aerodynamic model (solid) and reference
(dashed).

2.5 Linearized Aerodynamic Models

The aerodynamic models presented have in common that

■   a large number of parameters is required. These parameters may to some ex-
tent be confidential and not available for users of the model

■   the parameters give no insight into the physical relations of the model

■   the model is highly nonlinear, the calculation of initial values for the simulation is therefore possible only through iteration

The possible advantages of a linearized aerodynamic model are

■   a reduction of the number of parameters required

■   the use of parameters that show clear physical relations

■   a direct calculation of the initial values

■   reduction of the simulation effort

An approach for a linearized aerodynamic model had been proposed in [4],
the results had been compared to simulations with a more detailed model in
[21].But applicability of this model is limited to transmission system faults
cleared within 150 ms to 200 ms and the assumption of constant wind speed [22].

An improved model is proposed that includes the impact of changes of rotor
speed and changes of wind speed. As a result, the accuracy of the model for the

analysis of grid faults is increased and the range of applicability extended to the analysis of events like a voltage collapse with a longer time frame of interest [7].

2.5.1 Linearization along the Operation Trajectory

The aerodynamic power of a wind turbine as described by (2.15) and by substituting the tip speed ratio λ using (2.14) is a function of the independent variables blade angle Θ, rotor speed Ω_{WR} and wind speed v_W. A Taylor series

$$f(x) = f(a) + \frac{f'(a)}{1!}(x-a) + \frac{f''(a)}{2!}(x-a)^2 + \ldots \tag{2.23}$$

of a function with three variables and limited to the first derivative leads to

$$
\begin{aligned}
f(x,y,z) &\approx \\
&f(a,b,c) + f_x'(a,b,c)(x-a) + f_y'(a,b,c)(y-b) + f_z'(a,b,c)(z-c)
\end{aligned}
\tag{2.24}
$$

where f(a,b,c) denotes a calculated point with f'_x, f'_y, f'_z as the partial derivatives with respect to x, y and z at this point.

The aerodynamic power can then be written as:

$$
\begin{aligned}
P^*_{Aero}\left(\Theta, \Omega_{WR}, v_w\right) &\approx P^*_{Aero}\left(\Theta_0, \Omega_{WR,0}, v_{w,0}\right) \\
&+ \left.\frac{\partial P^*_{Aero}}{\partial \Theta}\right|_{(\Theta_0, \Omega_{WR,0}, v_{w,0})} \Delta\Theta + \left.\frac{\partial P^*_{Aero}}{\partial \Omega_{WR}}\right|_{(\Theta_0, \Omega_{WR,0}, v_{w,0})} \Delta\omega + \left.\frac{\partial P^*_{Aero}}{\partial v_w}\right|_{(\Theta_0, \Omega_{WR,0}, v_{w,0})} \Delta v_w
\end{aligned}
$$
$$\tag{2.25}$$

for a given operating point defined by Θ_0, $\Omega_{WR,0}$ and $v_{W,0}$. The deviation from the operating point is described by $\Delta\Theta = \Theta_0 - \Theta$, $\Delta\Omega_{WR} = \Omega_{WR,0} - \Omega_{WR}$ and $\Delta v_W = v_{W,0} - v_W$ [2].

The wind speed is usually assumed to be constant during the simulation (see section 2.5), therefore the partial derivative with respect to wind can be neglected. The aerodynamic power calculation can then be written in the following short form:

$$P_{Aero} = P_0 \quad + \quad \Delta P_\Theta \quad + \quad \Delta P_{\Omega_{WR}} \tag{2.26}$$

Note: The operator ∂ will be used in the following section for both analytical and numerical partial derivatives.

The function of the aerodynamic power as described by the Taylor series (2.25) and the short form in (2.26) describes the value of the power in the vicinity of an operating point given by P_0 and the derivative terms ΔP_Θ and $\Delta P_{\Omega_{WR}}$. As

---

[2]   Please note that a different sign convention is used in the text for Δ compared to (2.23)

shown in Fig. 2.3 - Fig. 2.5, the steady state operating points of the turbine can be described by a trajectory. The intention of this part is to expand the Taylor series description in (2.25) and (2.26) in a way that steady state operating point and the corresponding derivatives can be described for each operating point of the trajectory.

2.5.2 Steady state operating trajectory

The steady state operating point in (2.26) is given by $P_0 = P_{\text{Aero}}\left(\Theta_0, \Omega_{\text{WR},0}, v_{\text{w},0}\right)$, P_0 is defined as function of three independent variables. But Fig. 2.3 - Fig. 2.5 show a steady state operation trajectory that links the values of $\Omega_{\text{WR}}, \Theta$ and v_{W}. It shall be shown now that the variables P, Ω_{WR}, Θ can all be represented as functions of wind speed v_{W}. This will allow the definition of an operating trajectory as function of wind speed and is the basis for the further extension of (2.26) into a function valid at all operating points of the steady state operation trajectory.

Calculation of Steady State Blade Pitch Angle

For a turbine operation at below rated wind speed, the blade angle Θ equals zero[3]. The stationary blade angle Θ_{Stdy} for different wind speeds (most of the operating range of the turbine) is shown in Fig. 2.9. Using a quadratic curve fit as proposed in [22], the blade angle can be calculated as function of wind speed.

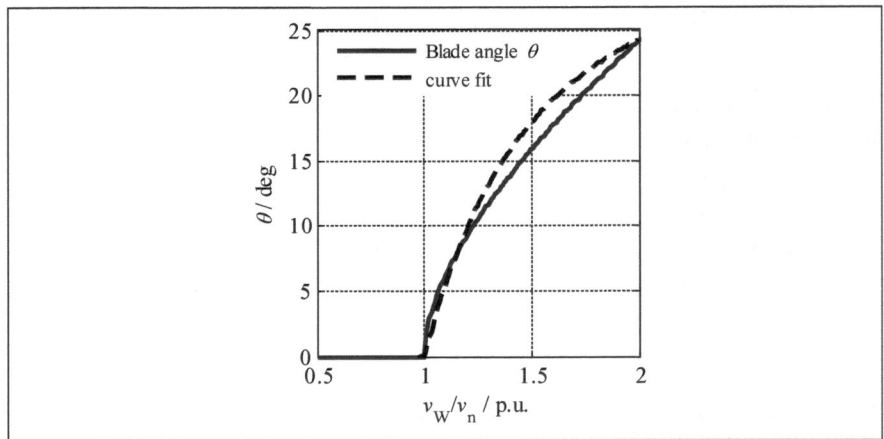

Fig. 2.9   Steady state blade angle trajectory as function of wind speed and approximation using a single parameter.

---

[3]   The small power increase due to an increase of blade angle at low wind speeds (see Fig. 2.3) can be ignored for most of the simulations.

Using a single parameter - the blade angle at twice the rated wind speed Θ_{VW2} - a good approximation can be achieved by

$$\Theta_{\mathrm{Stdy}}\left(v_{\mathrm{W}}^{*}\right)=\Theta_{\mathrm{VW2}}\frac{4}{3}\left(1-\frac{1}{v_{\mathrm{W}}^{*\,2}}\right) \tag{2.27}$$

with $v_{\mathrm{W}}^{*}=v_{\mathrm{W}}/v_{\mathrm{W,n}}$ as wind speed in p.u..

It can be shown that higher order fits with 2 or 3 parameters do not lead to an improved representation especially in the area usually of interest close to rated wind speed.

Calculation of Steady State Rotor Speed

As shown in Fig. 2.4, the wind turbine can only be operated at an optimal c_{P} - value in a limited operating range. For higher wind speeds, the tip speed ratio will decrease because the rotor speed has reached the allowed limit. From (2.14) follows that the steady state relation between steady wind speed and rotor speed is given (in p.u.) by

$$\omega_{\mathrm{WR,Stdy}}\left(v_{\mathrm{W}}^{*}\right)=\frac{\lambda_{\mathrm{opt}}v_{\mathrm{W,n}}}{R_{\mathrm{WR}}\Omega_{\mathrm{WR,n}}}v_{\mathrm{W}}^{*}=k_{\mathrm{\omega v}}v_{\mathrm{W}}^{*} \tag{2.28}$$

The rotor speed range is limited by the wind turbine limits. The steady state turbine speed as function of wind speed is shown in Fig. 2.10.

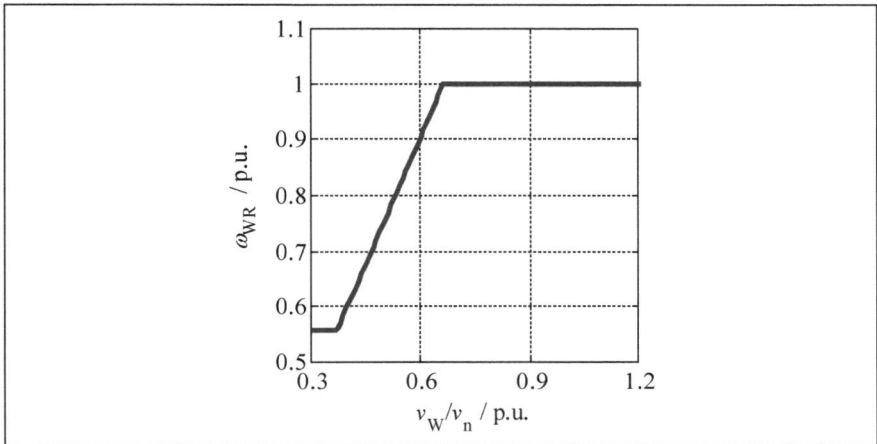

Fig. 2.10 Steady state turbine rotor speed as function of wind speed.

Calculation of Steady Turbine Power

As it can be seen from Fig. 2.4, the steady state aerodynamic power can be approximated in 3 steps.

(1) Below rated rotor speed, the turbine is operated at an optimal operating point, c_P can assumed to be constant. From Fig. 2.3 we see that this requires a constant tip speed ratio λ. As consequence of (2.28), the rotor speed is defined as function of wind speed. Eq. (2.15) describes a cubic relationship between wind speed and power for constant c_P until rated rotor speed is reached. The blade angle is kept constant.

(2) At rated rotor speed and higher wind speeds up to rated wind speed, a close to linear relation between wind speed and power (in the form $P = a \cdot v_W + b$) can be observed (see Fig. 2.11, also [22]). Rotor speed and blade angle are kept at constant values.

(3) For above rated wind speed, the aerodynamic power is assumed to be constant at 1 p.u..

The aerodynamic power of the steady state operation trajectory P_{Stdy} can therefore be approximated as function of wind speed only. The transition between the cubic part (1) and the linear part (2) of the power curve can be described by the power at the point rated rotor speed is reached (see Fig. 2.4). A variation of this value for different turbine types shows that the accuracy of the approximation decreases if the value is selected too high.

By selecting a value of $p = 0.3$ p.u. for the power it is ensured that the cubic part of the curve is limited for different turbine types. The transition point can be described as function of wind speed, in this case as the wind speed when a power output of $p{=}0.3$ p.u. is reached. This is the only parameter (called v_{WP03}) necessary to describe the steady state aerodynamic power.

The aerodynamic power is then described by

$$P_{Stdy}\left(v_W^*\right) = \begin{cases} \dfrac{0.3}{v_{WP03}^3} v_W^{*\,3} & \text{for } v_{W,min} < v_W^* < v_{WP03} \\[2ex] 1 + \left(v_W^* - 1\right)\dfrac{0.7}{1 - v_{WP03}} & \text{for } v_{WP03} < v_W^* < 1 \\[2ex] 1 & \text{for } v_W^* \geq 1 \end{cases} \qquad (2.29)$$

with v_{Wmin}^* as the minimum wind speed required for operation of the wind turbine, with all quantities in p.u. and v_n as rated wind speed. The aerodynamic power from a detailed calculation and using the approximation of (2.29) is shown in Fig. 2.11.

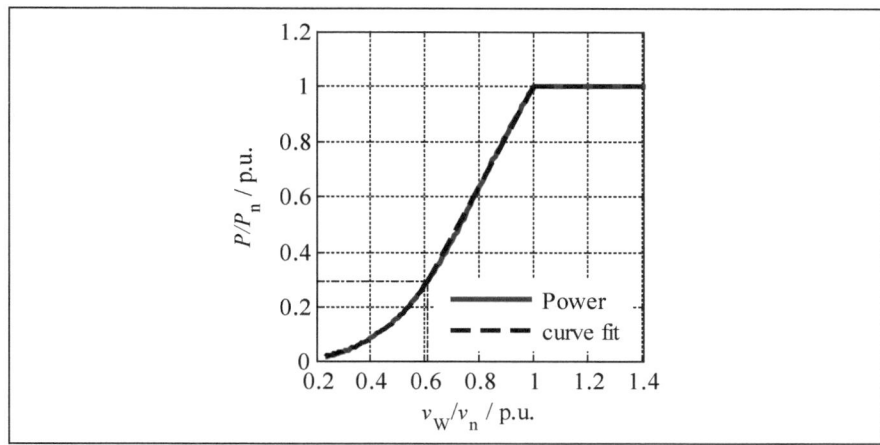

Fig. 2.11 Steady state turbine aerodynamic power as a function of wind speed and
approximation using a single parameter.

2.5.3 Partial Derivative ΔP_Θ: Change of Power with respect to Blade Pitch Angle

The partial derivative ΔP_Θ in (2.26) shall describe the impact of blade angle
changes on the power output in the vicinity of an operating point given by the
steady state operation trajectory. Starting point for the calculation is the representation of the aerodynamic power as function of blade angle and wind speed. It
can be calculated from the blade data shown in Fig. 2.5 using (2.15) and is shown
in Fig. 2.12 for a 2 MW wind turbine. The thick green line represents the steady
state operation trajectory showing the relation between blade angle, wind speed
and aerodynamic power under steady state conditions. Beyond rated wind speed,
the blade angle is increased to limit the aerodynamic power to 1 p.u..

Based on Fig. 2.12, the partial derivative of the aerodynamic power de-
scribed by (2.25) with respect to blade angle $\partial P / \partial \Theta$ has been calculated (see
Fig. 2.13). The thick green line represents the same steady state operation trajectory as shown in Fig. 2.12.

For the definition of the ΔP_Θ, only the partial derivatives along the operation
trajectory are relevant. The partial derivative $\partial P/\partial \Theta$ along the operation trajectory can be described as function of blade angle Θ (see Fig. 2.13), as function of
wind speed v_W (see Fig. 2.15) and as function of rotor speed Ω_{WR}. But since rotor
speed is constant for all blade angles >0, (see Fig. 2.4), this is not an option.

The derivative $\partial P/\partial \Theta$ should be described as function of blade angle. This allows a description by a simple linear equation, and a change of blade angle below
rated wind speed can also be represented (a blade angle of e.g. 10 deg at rated
wind speed could not be represented by the function in Fig. 2.15).

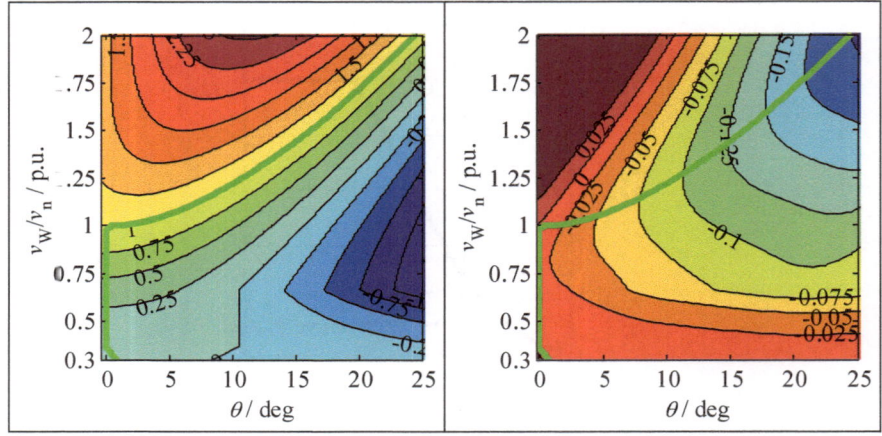

Fig. 2.12 Aerodynamic power in p.u. at the turbine rotor as function of blade angle Θ and wind speed v_W. The turbine is operated along the green steady state operation trajectory.

Fig. 2.13 Partial derivative $\partial p / \partial \Theta$ in p.u. as function of blade angle Θ and wind speed v_W. The green line represents the steady state operation trajectory of the turbine and corresponds with the green line in Fig. 2.12.

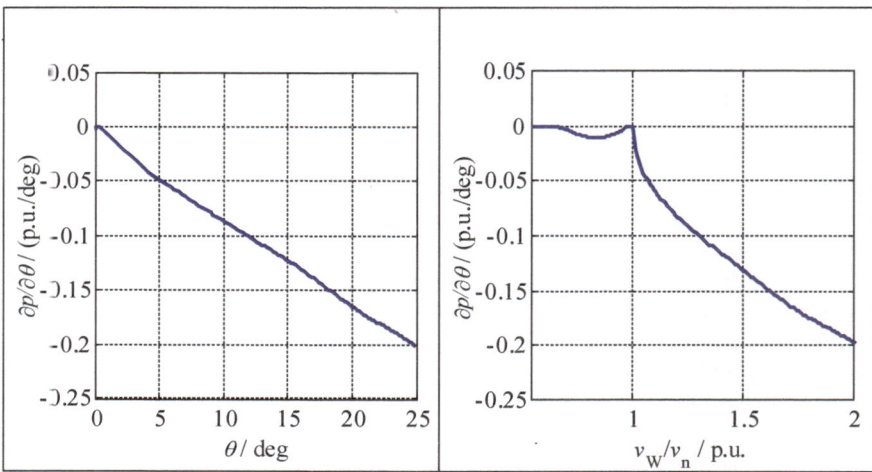

Fig. 2.14 Partial derivative $\partial p / \partial \Theta$ in p.u. along the steady state operation trajectory as function of blade angle Θ.

Fig. 2.15 Partial derivative $\partial p / \partial \Theta$ in p.u. along the steady state operation trajectory as function of wind speed v_W.

The first approach of linearizing $\partial P / \partial \Theta$ along the stead state operation trajectory calculates the derivatives by using small changes of ± 0.1 deg around the operation trajectory (see blue line in Fig. 2.17). But the analysis of typical definitions of grid fault duration and remaining voltage level in grid codes shows, that voltage profiles are usually either "short and deep" for a nearby fault, or longer but less severe for remote faults [23], [24]. During such grid faults, a typical change of the blade angle by no more than +5 deg can be observed in measurements and simulations.

By using a larger change of the blade angle of +5 deg for the calculation of the derivative, an improved accuracy of the simulation results for typical grid faults can be achieved (green line in Fig. 2.17). The blue line represents the linearization with equal positive and negative difference which represents the derivative at the operation trajectory (see also point P_1 in Fig. 2.16). The green line represents the average of the derivatives at the operation trajectory and for a point with a difference in blade angle of +5 deg (point P_2 in Fig. 2.16). The results are shown in Fig. 2.17 by the green line; the black dashed lines represent linearized curves obtained by curve fit.

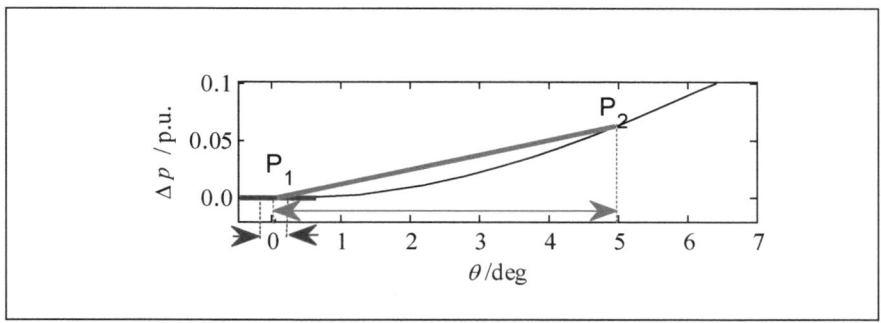

Fig. 2.16 Linearization of power change for typical blade angle changes.

The resulting function $\Delta P_\Theta = f(\Theta)$ can be described as a simple linear equation:

$$\Delta P_\Theta = \left(\Theta_{\text{Stdy}} - \Theta \right)\left(K_{\text{Aero}} \cdot \Theta_{\text{Stdy}} + C_{\text{Aero}} \right) \qquad (2.30)$$

with Θ_{Stdy} defined by (2.27). Typical parameter ranges are $k_{\text{Aero}} = -0.007 \; .. \; -0.008 \; \text{p.u./deg}^2$ and $c_{\text{Aero}} = -0.025 \; .. \; -0.035 \; \text{p.u./deg}$ for wind turbines in the range of 2 MW .. 6 MW (for a calculation in p.u. values). This function describes the impact of blade pitch angle changes on the turbine power for the entire steady state operation range.

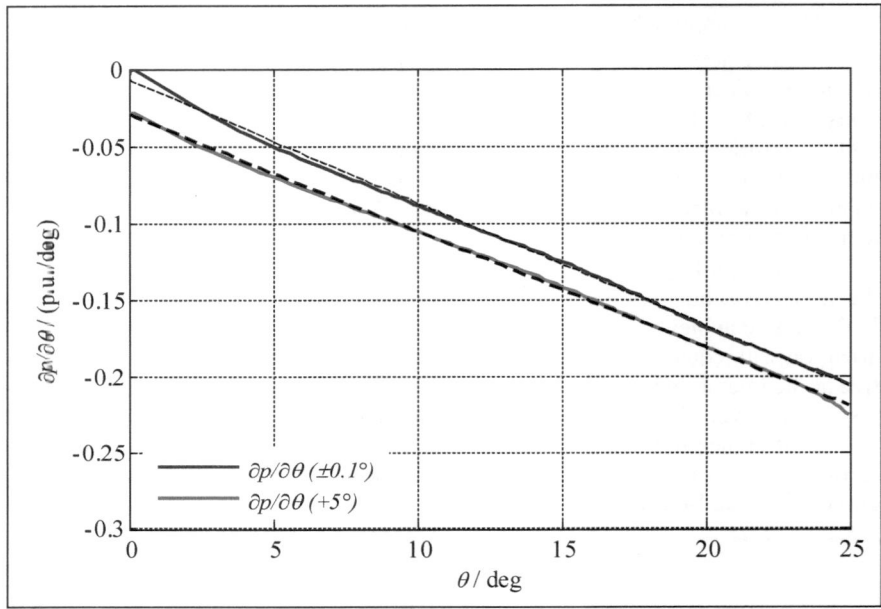

Fig. 2.17 Rate of change of power due to changes of blade angle $\partial P / \partial \Theta$ for small changes
(blue) and large changes (green) around the steady state operation trajectory.

An implementation of the partial derivative ΔP_Θ , change of power with respect to blade angle ($\partial P / \partial \Theta$) along the operation trajectory is shown in Fig.
2.18.

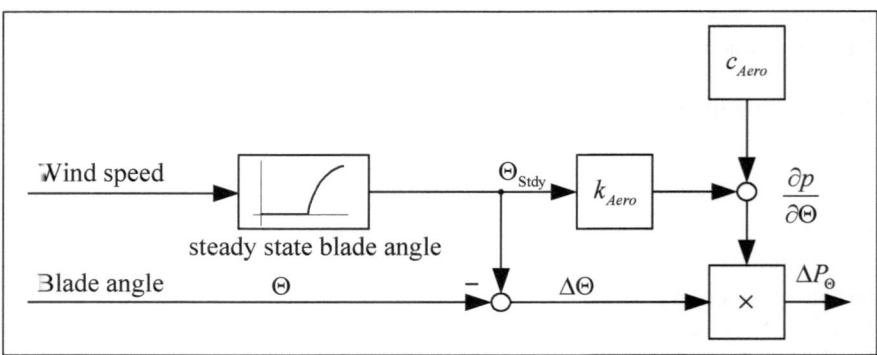

Fig. 2.18 Model implementation of partial derivative ΔP_Θ : change of power with respect to
blade angle ($\partial P / \partial \Theta$).

2.5.4 Partial Derivative $\Delta P_{\Omega,\mathrm{WR}}$: Change of Power with respect to Rotor Speed

The partial derivative $\Delta P_{\Omega,\mathrm{WR}}$ in (2.26) shall describe the impact of rotor speed changes on the power output in the vicinity of an operating point given by the steady state operation trajectory. Starting point for the calculation is the representation of the aerodynamic power as function of blade angle and wind speed as shown in Fig. 2.12 in the previous section.

Based on this figure, the partial derivative of the aerodynamic power with respect to rotor speed $\partial P / \partial \Omega_{\mathrm{WR}}$ has been calculated (see Fig. 2.24). The thick green line represents the same steady state operation trajectory as shown in Fig. 2.12.

For the definition of $\Delta P_{\Omega,\mathrm{WR}}$ only the partial derivatives along the operation trajectory are relevant. The partial derivative $\partial P / \partial \Omega_{\mathrm{WR}}$ along the operation trajectory can be described as function of blade angle Θ (see Fig. 2.20) as function of wind speed v_W (see Fig. 2.21) and as function of rotor speed Ω_{WR}. But the rotor speed is constant for all blade angles >0, (see Fig. 2.4), therefore describing $\partial P / \partial \Omega_{\mathrm{WR}}$ as function of rotor speed would not provide useful results.

The linearization of $\partial P / \partial \Omega_{\mathrm{WR}}$ should be described as function of wind speed (Fig. 2.21), since the effect of changes of the rotor speed can also be relevant below rated wind speed if more energy is extracted from the turbine than the wind conditions allow following frequency deviations [25]. In this case, the turbine is forced into an aerodynamic operating point that differs from the optimal trajectory. A representation as function of blade angle (see Fig. 2.20) would not provide results below rated wind speed where the blade angle is constant.

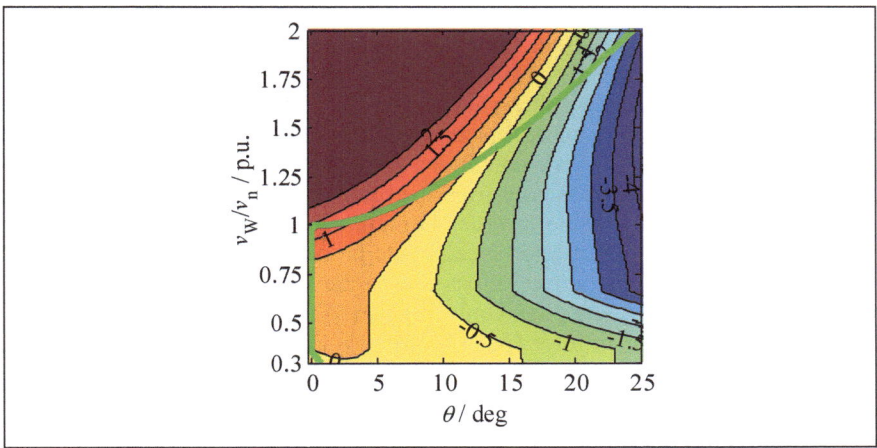

Fig. 2.19 Partial derivative $\partial p / \partial \, \Omega_{\mathrm{WR}}$ in p.u. as function of blade angle and wind speed. The green line represents the steady state operation trajectory of the turbine and corresponds with the green line in Fig. 2.12.

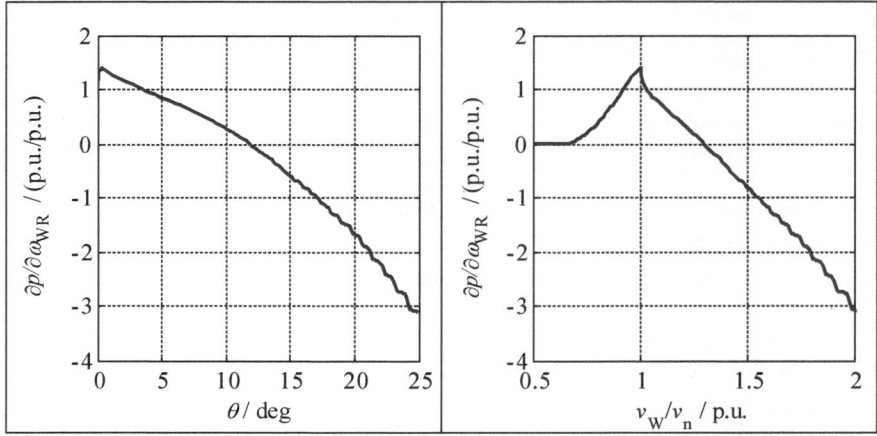

Fig. 2.20 Partial derivative $\partial p_{\text{Aero}}/\partial \omega_{\text{WR}}$ in p.u. along the steady state operation trajectory as function of blade angle Θ.

Fig. 2.21 Partial derivative $\partial p_{\text{Aero}}/\partial \omega_{\text{WR}}$ in p.u. along the steady state operation trajectory as function of wind speed v_{W}.

The change of power with respect to rotor speed changes of 0.01 p.u. is shown in the blue line in Fig. 2.22. The function $\Delta P_{\Omega,\text{WR}}$ can be represented by

$$\Delta P_{\Omega,\text{WR}} = \left(\Omega_{\text{WR,Stdy}} - \Omega_{\text{WR,}} \right) \cdot f'_{\Omega,\text{WR}} \left(v_{\text{W}} \right) \qquad (2.31)$$

with $f_{\Omega}\left(v_{\text{W}}\right)$ as partial derivative $\partial P/\partial\Omega_{\text{WR}}$ as function of wind speed. A fast increase of rotor speed can result from a grid fault where the power output of the turbine is reduced suddenly while the aerodynamic power stays constant or changes slowly only due to a slow change of blade angle.

A decrease of turbine speed can also result, if more energy is extracted from the turbine than the actual wind conditions allow. This may be the case following grid disturbances if kinetic energy from the rotor is used to temporarily increase the power output of a turbine in order to minimize frequency deviations in the grid [25]. These changes of rotor speed could be considerably larger than speed excursions due to voltage dips. This is due to the fact that an increase in rotor speed can be limited by the pitch controller curtailing the power captured by the wind, while a decrease in rotor speed is due to a lack for energy that cannot be compensated. How much the rotor speed decreases in such a case depends only on the turbine controller settings.

Representing $f'_{\Omega,\text{WR}}\left(v_{\text{W}}\right)$ as function of wind speed shows three different areas in Fig. 2.22:

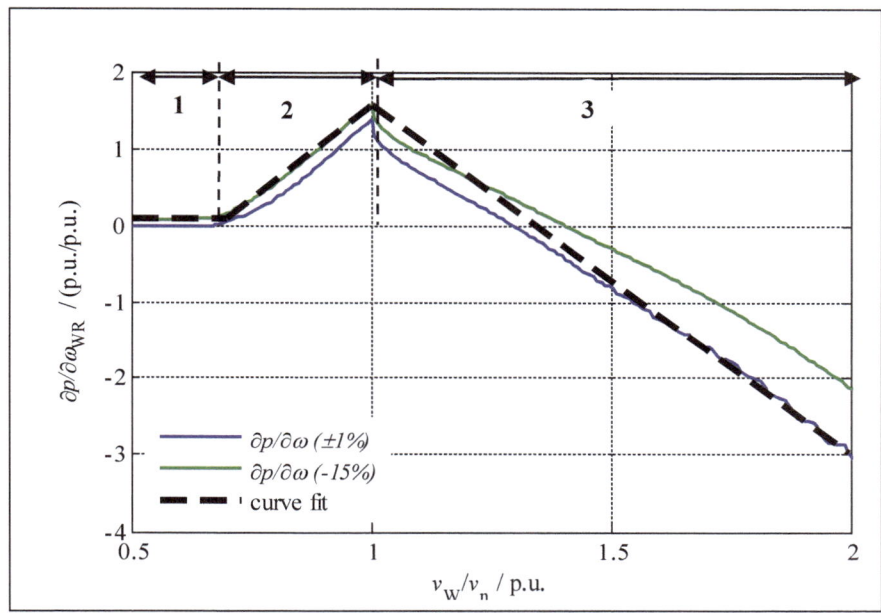

Fig. 2.22 Rate of change of power due to changes of rotor speed for small changes (blue) and large changes (green) around the operation trajectory.

(1) from cut-in wind speed to rated rotor speed (at $v_W \approx 0.7$ p.u., see also Fig. 2.4): $\partial p/\partial \omega_{WR}$ is rather small and almost constant. The rotor blade is operated near its optimal operating point. Changes in rotor speed lead to small or no changes in blade angle.

(2) from rated rotor speed to rated wind speed ($v_W = 1$ p.u.): $\partial p/\partial \omega_{WR}$ increases with increasing power output. This is the area most relevant for kinetic energy support. A very good linearization is possible in this area. Here again, no or only minor changes of the blade angle can be expected.

(3) above rated wind speed ($v_W > 1$ p.u.): $\partial p/\partial \omega_{WR}$ decreases. A curve fit leads to somewhat larger deviations from the derivative function, but this is not critical because differences between curve fit and derivative function curve will only lead to deviations in the rotor speed dynamic but do not have an impact on power output.
At above rated wind speed, only a small increase in power output can be expected since the turbine rating would otherwise be exceeded. Power will usually not increase by more than 5% [25].

The increase in power output will initially lead to a reduction of rotor speed, which will cause the pitch-speed controller to decrease the blade angle in order to return the turbine speed back to rated speed. This is possible because the wind potential (at above rated wind speed) is higher than the turbine power output. The proposed curve fit along the blue line (small changes of rotor speed) for higher wind speeds is therefore reasonable.

Implementation of $\Delta P_{\Omega,\text{WR}}$

The effect of changing rotor speed $\partial P/\partial \Omega_{\text{WR}}$ can therefore be described as function of wind speed $P_{\Omega} = f(v_{\text{W}})$ consisting of a combination of three lines. A simple implementation is possible using a table with only four parameters (all in p.u.) describing the wind speed ($v_{\omega 1}^{*}$) and power change $\partial p_{\omega 1}$ when rated rotor speed is reached, the rate of change at rated $\partial p_{\text{V1}}^{*}$ and at twice the rated wind speed $\partial p_{\text{V2}}^{*}$

Using the table implementation, the value $\partial p/\partial \omega_{\text{WR}}$ in (p.u.) for each value of wind speed v_{W}^{*} can be calculated using a linear interpolation with

$$
\begin{array}{llllll}
input\ \left(v_{\text{W}}^{*}\right) & : & 0 & v_{\omega 1}^{*} & 1 & 2 \\
output\ \left(\partial p/\partial \omega_{\text{WR}}\right) & : & \partial p_{\omega 1} & \partial p_{\omega 1} & \partial p_{\text{V1}}^{*} & \partial p_{\text{V2}}^{*}
\end{array}
\tag{2.32}
$$

Typical values are: $v_{\omega 1}^{*} = 0.35 .. 0.4$ p.u., $\partial p_{\omega 1} = 0.5 .. 0.7$ p.u., $\partial p_{\text{V1}}^{*} = 1.3 .. 1.6$ p.u. and $\partial p_{\text{V2}}^{*} = -2.5 .. -3.5$ p.u..

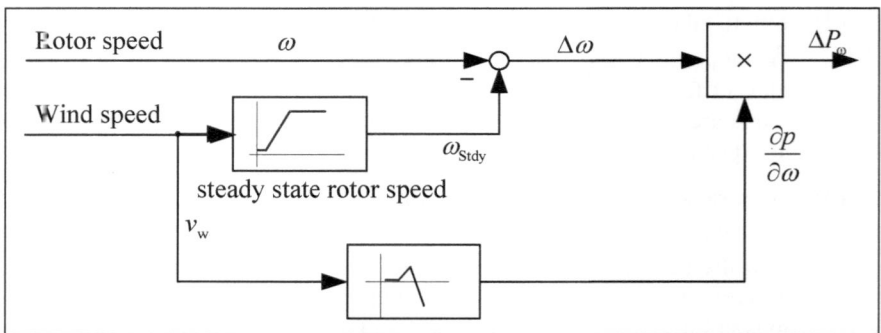

Fig. 2.23 Model implementation of partial derivative $\Delta P_{\Omega,\text{WR}}$: change of power with respect to rotor speed ($\partial P/\partial \Omega_{\text{WR}}$).

An implementation of the partial derivative $\Delta P_{\Omega,\text{WR}}$, change of power with respect to rotor speed change ($\partial P/\partial \Omega_{\text{WR}}$), along the operation trajectory is shown in Fig. 2.23.

Relevance of $\partial P/\partial \Omega_{\text{WR}}$

In [21] it was assumed that the effect of rotor positive (that means accelerating) speed changes as a result of grid faults was not relevant for grid faults of less than 200 ms duration. But for grid faults described in [23], [24] and the test defined in [26], speed excursions of up to 0.05 p.u. can be expected. For the blade described in Fig. 2.5 with a maximum $\partial p/\partial \omega_{\text{WR}}$ of 1.4 p.u. for small changes of rotor speed, this leads to a change of up to 7% of rated power:

$$\Delta p_\Omega \left(v_w^* = 1 \right) = \left(\partial p / \partial \omega_{\text{WR}} \right)_{\text{max}} \cdot \left(\Delta \omega_{\text{WR}} \right)_{\text{max}} = 1.4 \cdot 0.05 = 0.07 \qquad (2.33)$$

This is between 10% and 15% of the total change ΔP_{Aero} during grid faults (see Fig. 2.31).

2.5.5 Resulting model structure

The complete aerodynamic model is shown in Fig. 2.24, consisting of the calculation of the steady state aerodynamic power p_{Stdy} and the derivative terms of power change with respect to blade angle ΔP_Θ and power change with respect to rotor speed $\Delta P_{\Omega,\text{WR}}$. The blocks highlighted in blue can be moved into an initial value calculation if constant wind speed is assumed throughout the simulation. The simplified structure with constant wind it is used for example in the standard IEC 61400-27 [27].

2.5.6 Calculation initial Values for Wind Speed

Commonly an initial value for the output power at the turbine terminals is given from the load flow calculation. Based on this value, the initial power at turbine rotor ($P_{\text{Aero},0}$) can be calculated if turbine losses are neglected. For a simplified generic model, it is recommended to include the turbine losses in the aerodynamic model. In the next step, an equivalent wind speed needs to be calculated. Using power as input, (2.29) can be transformed to calculate the initial wind speed v_w^* (in p.u.) for a given initial aerodynamic power $p_{0\text{Aero}}$ (in p.u.) in (2.34).

$$v_w^* = 1 + \frac{1 - v_{\text{WP03}}}{0.7} \left(P_{\text{Aero},0} - 1 \right) \qquad (2.34)$$

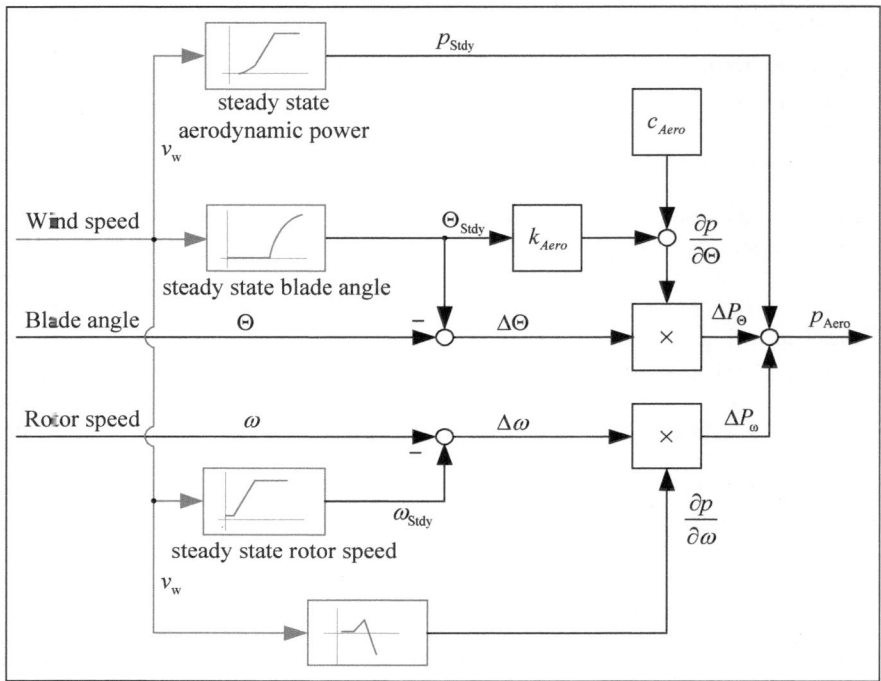

Fig. 2.24 Implementation of the aerodynamic model. If constant wind speed is assumed, the blocks highlighted in blue can be moved into an initial value calculation which reduces the complexity of the dynamic model.

2.5.7 Operation of the Turbine with Active Power Limitation

If the turbine is operated at an active power below its optimal operating point for wind speeds < 1 p.u., different operating trajectories need to be considered (see Fig. 2.25). The linearization of the partial derivatives $\partial P/\partial\Theta$ for rated power and power limitation to 0.5 p.u. is shown in Fig. 2.26 (the thin blue and green lines describe the operation at reduced power). As it can be seen, the change of blade angle ($\partial P/\partial\Theta_{WR}$) is not very sensitive to the active power level. A limitation of the active power to 0.5 p.u. leads only to limited changes in the partial derivative.

By contrast, the impact of rotor speed changes $\partial P/\partial\Omega_{WR}$ (see Fig. 2.27) changes considerably if the active power is limited to 0.5 p.u. In the case of large blade angle changes due to a reduction of the turbine active power reference, large blade angle changes have to be expected. The definition of $\partial P/\partial\Omega_{WR}$ as function of wind speed as shown in section 2.5.4 is not an acceptable solution in this case; an extension of the aerodynamic model is necessary.

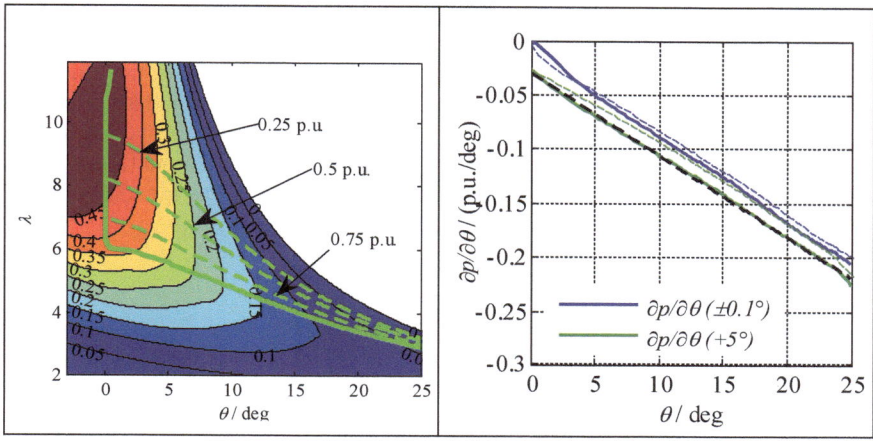

Fig. 2.25 Turbine operation trajectories for active power limitation to 0.25 p.u., 0.5 p.u., 0.75 p.u. and 1 p.u. .

Fig. 2.26 Rate of change of power due to changes of blade angle $\partial P/\partial\Theta$ for 1 p.u. and 0.5 p.u. active power (thin lines).

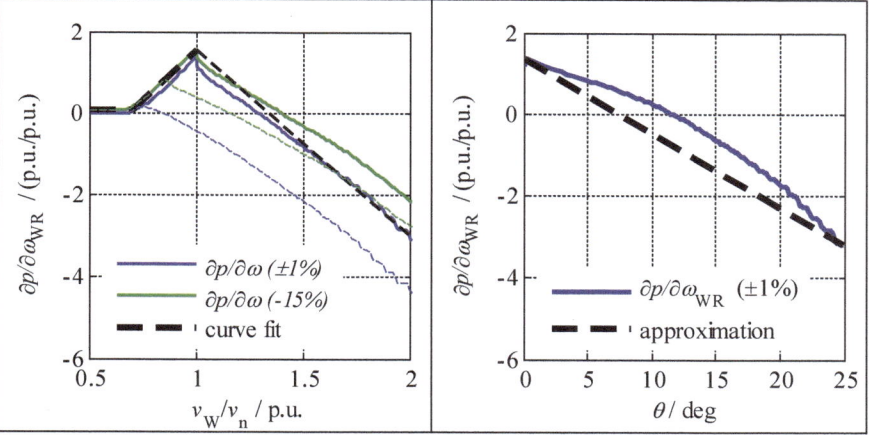

Fig. 2.27 Rate of change of power due to changes of rotor speed $\partial P/\partial\Omega_{WR}$ for 1 p.u. (solid line) and 0.5 p.u. (dashed line) active power as function of wind speed.

Fig. 2.28 Rate of change of power due to changes of rotor speed $\partial P/\partial\Omega_{WR}$ for 1 p.u. active power as function of blade angle.

Modification of the Impact of $\partial P/\partial\Omega_{WR}$

For wind speeds up to rated and no active power curtailment, the blade angle is assumed to be zero (see Fig. 2.9). For wind speeds higher than rated and for active power limitations, the blade angle will change. An improved representation of large blade angle changes without the need for additional parameters can be achieved by combining the characteristics of $\partial P/\partial\Omega_{WR}$ as function of wind speed (Fig. 2.27) and of $\partial P/\partial\Omega$ as function of blade angle Θ (Fig. 2.28)

This is achieved by creating a modified function

$$\partial P/\partial\Omega = f(v_w) - f(\Theta_0 = 0) + f(\Theta) \tag{2.35}$$

and using a modified parameterization for $f(v_w)$ that stay at a constant value for $v^*_W > 1$.

$$\partial P/\partial\Omega = \begin{cases} f(v_w) & \text{for } v_w \leq 1 \text{ and } \Theta = 0 \text{ (partial load, no curtailment)} \\ f(v_w) + f(\Theta) & \text{for } v_w \leq 1 \text{ and } \Theta > 0 \text{ (partial load or curtailment)} \\ f(v_w) + f(\Theta) & \text{for } v_w > 1 \text{ and } \Theta > 0 \text{ (full load or curtailment)} \end{cases} \tag{2.36}$$

The resulting model representation describing the impact of rotor speed changes on $\partial P/\partial\Theta$ the power output of the turbine rotor is described in Fig. 2.29.

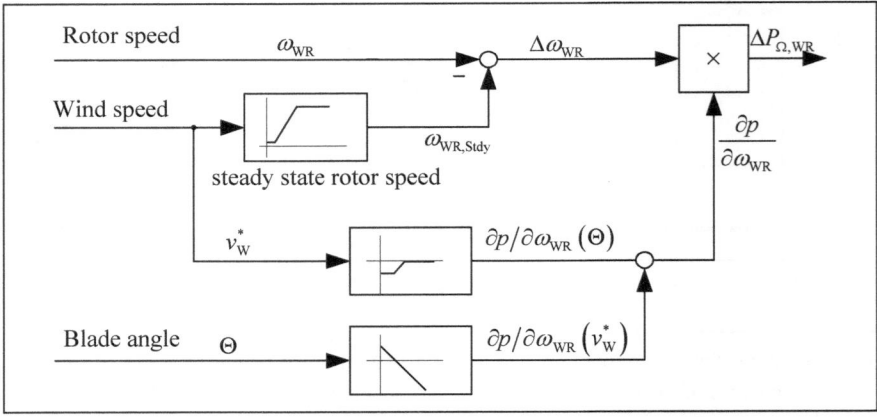

Fig. 2.29 Model implementation of partial derivative $\Delta P_{\Omega,WR}$: change of power with respect to blade angle ($\partial P/\partial\Theta$) with extension for large blade angle deviations.

Initial Value Calculation

For initial conditions of the simulation using an active power limitation, that means for conditions where wind speed would allow a higher power output than currently demanded, a blade pitch angle greater zero needs to be calculated. An example would be an operation at rated wind speed, but with an aerodynamic power P_{Aero} of less than rated power. By setting the initial value $P_{\text{Aero},0} = P_{\text{Aero}}$, and inserting (2.30) in (2.26) the initial blade angle is

$$\Theta_0 = \Theta_{\text{Stdy}} - \frac{P_{\text{Aero},0} - P_{\text{Stdy}} - \Delta P_{\Omega,\text{WR}}}{K_{\text{Aero}} \cdot \Theta_{\text{Stdy}} + C_{\text{Aero}}} \qquad (2.37)$$

with the steady state values blade angle Θ_{Stdy} given by (2.27), and the aerodynamic power given by (2.29). The term $\Delta P_{\Omega,\text{WR}}$ would normally be zero unless an initialization at a non-steady-state rotor speed was intended, for example to validate a given measurement where the turbine at the start of the measurement is not operating exactly at its steady state value.

The resulting structure of the aerodynamic model is shown in Fig. 2.30. The blocks highlighted in blue can be moved into an initial value calculation if constant wind speed is assumed throughout the simulation.

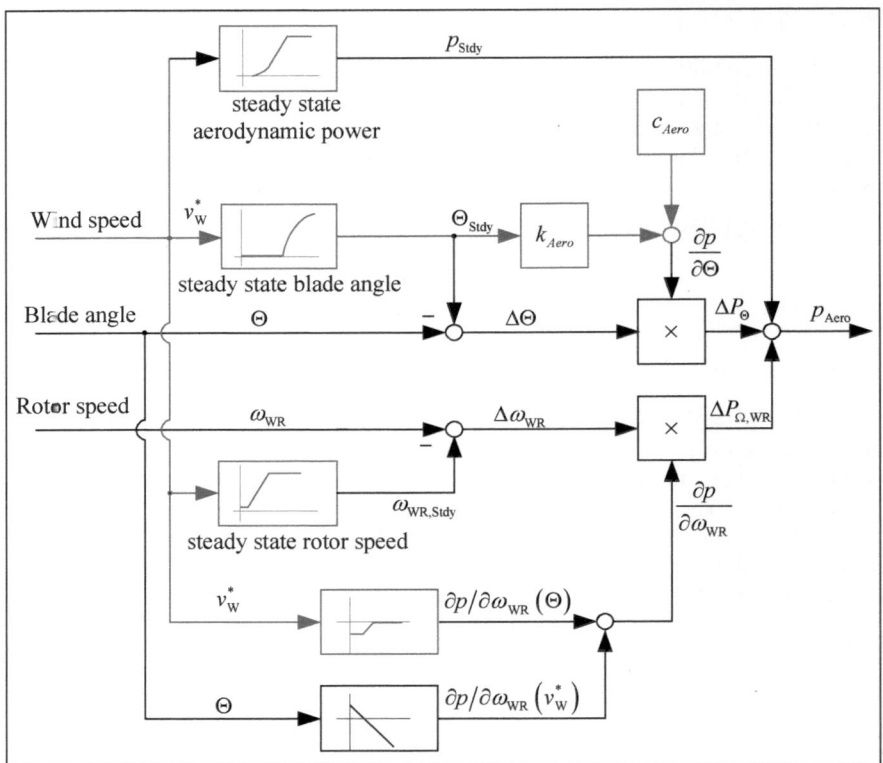

Fig. 2.30 Structure of the aerodynamic model with the three components (1) steady state aerodynamic power, (2) impact of blade angle change and (3) impact of rotor speed change.

2.5.8 Comparison with c_P-λ Table Representation

Simulations using the proposed aerodynamic model are compared to a c_P-λ table. The simulation of a voltage dip down to 0% for 150 ms using the proposed model and as comparison using a c_P-λ table as aerodynamic model is shown in Fig. 2.31. Even though there are some differences in the calculated aerodynamic power, the results for turbine speed and power are very close to each other.

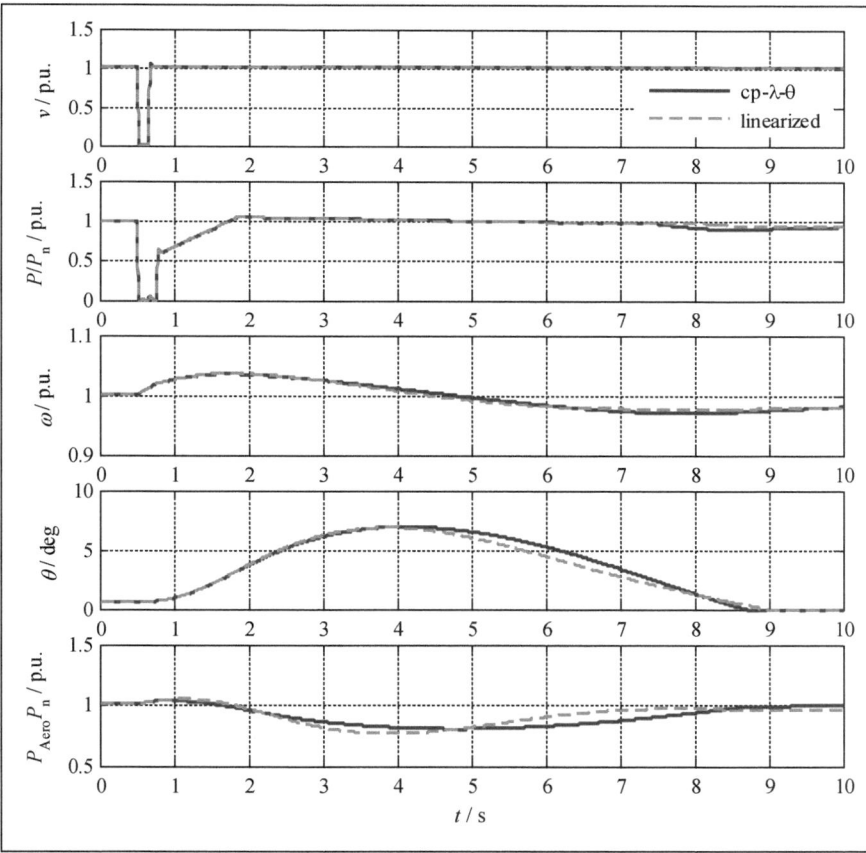

Fig. 2.31 Comparison of aerodynamic models: Voltage drop down to 0% for 150 ms using c_P-λ (solid) and linearized (dashed) aerodynamic model.

The turbine response to a change of wind speed using both the proposed model and a c_P-λ table is shown in Fig. 2.32. As it can be expected from Fig. 2.9, there is a difference between the blade angles of up to 3 deg at blade angles between 15 deg and 20 deg. Still, the turbine speed and the power output are almost identical for both aerodynamic models. If a variable wind speed is used, as with c_P-λ tables, a first order lag of 2 s .. 4 s should be used to cancel out higher frequency components in the wind speed [28].

Both the linearized approach and c_P-λ tables are usually not accurate enough to simulate fast changes of wind speed during gusts (see section 2.6). For a detailed analysis of such phenomena, aerodynamic models based on blade element theory are recommended.

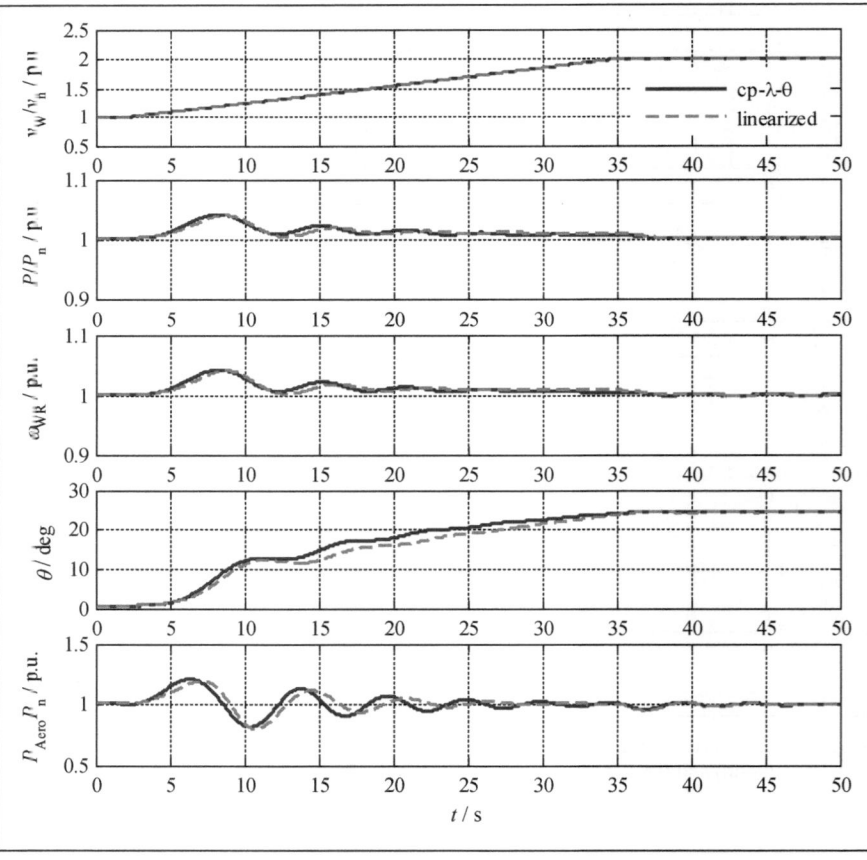

Fig. 2.32 Comparison of aerodynamic models: Change of wind speed using c_P-λ (solid) and linearized (dashed) aerodynamic model.

2.5.9 Comparison with other c_P-λ Representations

A comparison of the c_P-λ curves for different representations has been shown in Fig. 2.6 - Fig. 2.8 in section 2.4. From these diagrams, differences are visible, but the impact for simulations is not evident. Therefore, the partial derivatives of changes of power with respect to blade angle $\partial p/\partial \Theta$ and rotor speed $\partial p/\partial \omega_{WR}$ of different aerodynamic models are compared in Fig. 2.33 and Fig. 2.34.

The original functional representation in (2.16)-(2.18) by [18] (dash-doted curve) shows large differences compared to the c_P-λ table that has been chosen as reference. The modified functional representation (dashed curve in the in Fig. 2.33 and Fig. 2.34) given by (2.19) and (2.20) shows an improved representation. Compared to these results, the polynomial representation given by (2.21) and (2.22) (red curve) offers a significant improvement.

Even though good results can be achieved using a polynomial approach, there remain drawbacks. First, a numerical optimization is necessary to calculate parameters that fit well. If this is omitted, more parameters will be needed for a good estimate. Especially the polynomial approach is computationally intensive if a new value is calculated at ever simulation step. If, as an alternative, a table is calculated during initialization, either a high number of table entries are needed or a higher order interpolation between table entries is necessary, which again increases model complexity and computational requirements. In addition to the c_P-λ representation, additional turbine parameters (rotor diameter, gearbox ratio, losses ...) are required for this approach.

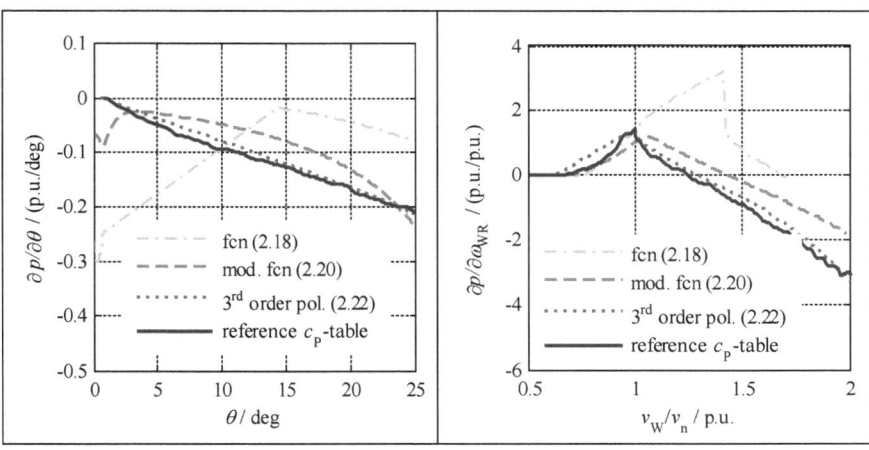

Fig. 2.33 Comparison of $\partial p/\partial \Theta$ along the steady state operation trajectory for different c_P-λ representations as function of blade angle.

Fig. 2.34 Comparison of $\partial p/\partial \omega_{WR}$ along the steady state operation trajectory for different c_P-λ representations as function of wind speed.

2.6 Representation of Dynamic Inflow Phenomena

Up to this point, a static representation of the rotor has been used. The analysis of wind turbines using detailed aerodynamic models [9] shows that fast changes of the blade angle can lead to transient changes of power. These models are based on detailed blade element theory models. Fast changes of blade angle especially around rated wind speed may not be represented accurately [9].

An approach to represent dynamic effects in a simplified model was presented by [29], but this approach is still based on static c_P-λ tables. If such effects are relevant, a more detailed representation of the rotor aerodynamics is necessary Simplified models to describe this effect have been proposed in [30] and [31]. the application for variable speed wind turbine models has been discussed in [10], [32] and [33].

Including dynamic inflow phenomena into a generic aerodynamic model could lead to a slight increase in modeling accuracy of the aerodynamic model. But this would be at the expense of a higher number of parameters for a general solution. A reduced number of parameters would be needed for a solution tuned to a specific wind speed. Case-specific optimization, this means modifying the parameters of the aerodynamic model for a specific wind speed, could improve accuracy as well without an increase of parameters. But such an approach as proposed in [27] may only be accepted in a limited number of cases.

For model validation, inaccuracies of the wind speed measurement and the impact of wind speed changes during the measurement have far more impact on the results than further improvements in the aerodynamic model.

2.7 Comparison with Measurements

The comparison of aerodynamic models with measurements of wind turbines is clearly limited in its accuracy by the fact that the wind speed as input can only be measured with a high degree of uncertainty. Still, an event like a grid fault triggers reproducible results if the change of wind speed during and following the grid fault is limited.

In Fig. 2.35 a voltage dip with a measurement setup according to Section A.4. Fig. A.3 was used to create a voltage dip down to around 50% rated voltage for 2 s. A 2 MW turbine was operating at rated power with a wind speed at about rated. The turbine simulation model is based on the generic WECC/IEEE type 3 model [4], with an extended generator model according to [34] and section 5. It uses a two mass representation of the drive train according to section 3.3.2 and a control structure according to section 4. The wind speed is kept constant during the simulation.

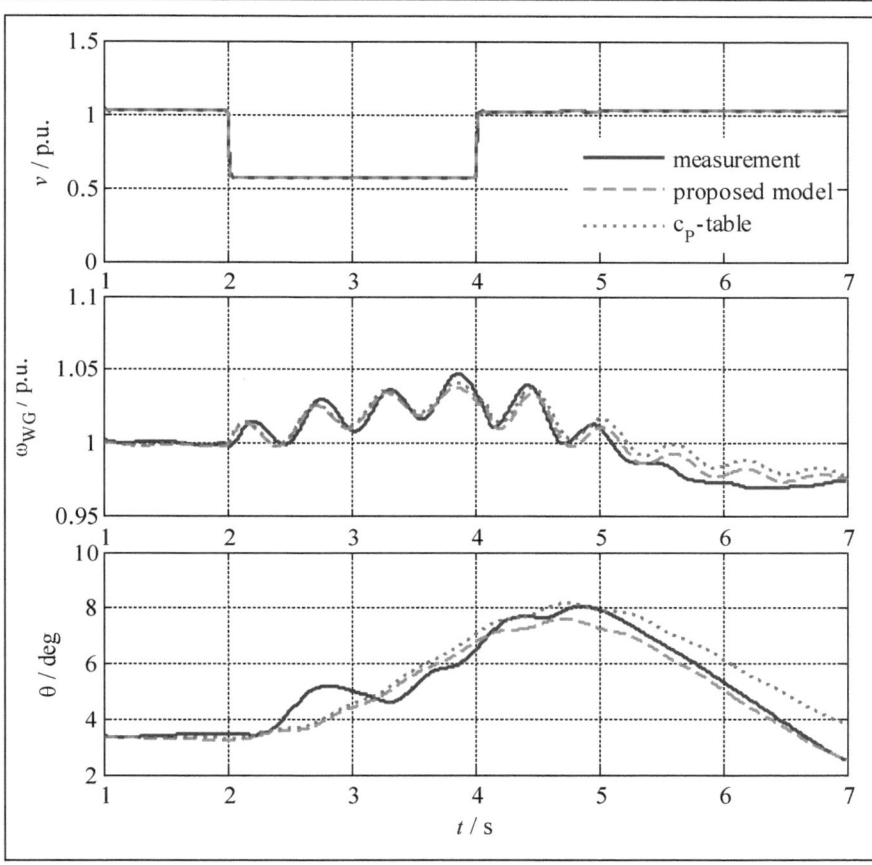

Fig. 2.35 Comparison of measurement of a 2 MW turbine (solid), simulation using the proposed aerodynamic model (dashed) and simulation using c_P-λ table (doted) for a voltage dip down to 50% rated voltage at rated power.

The measurement is compared to simulations using (a) the proposed aerodynamic model and (b) an aerodynamic model based on a static c_P-λ table. Even though the real control structure of the turbine has been replaced by a relatively simple simulation model, the simulation results are relatively close to measurement. The difference between the model using a c_P-λ table and the proposed model is small. Both blade angle and rotor speed are close to measurement. The oscillation visible in the rotor speed is due to the first drive train eigenfrequency (see section 3.3.2 for more details).

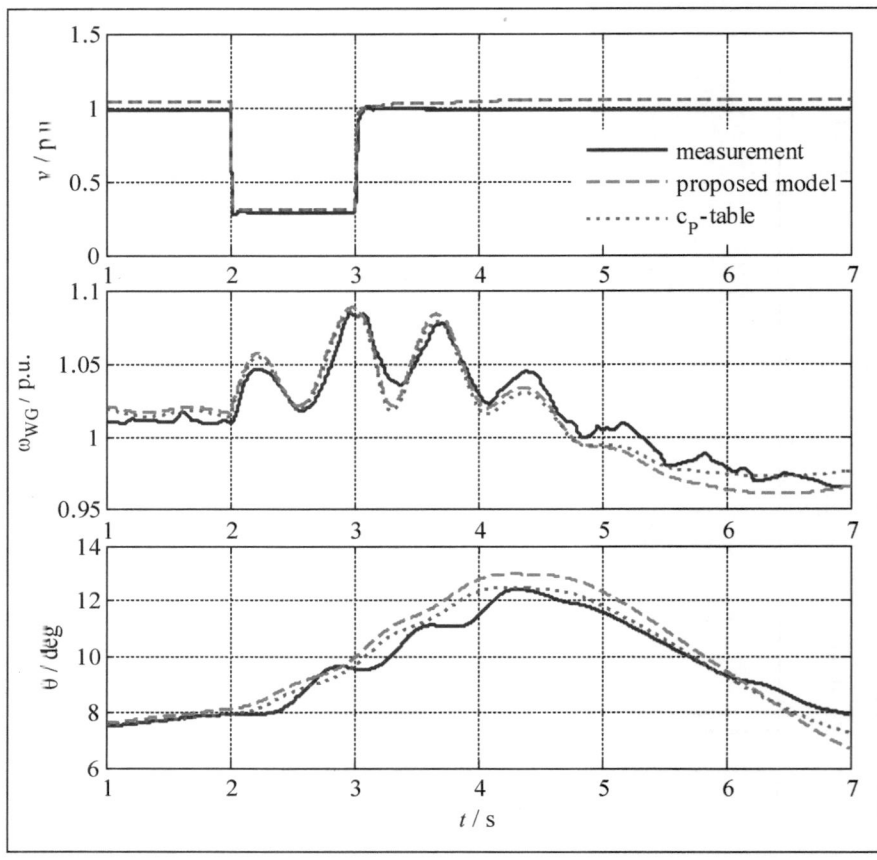

Fig 2.36 Comparison of measurement of a 6 MW turbine (solid), simulation using the
proposed aerodynamic model (dashed) and simulation using c_P-λ table(doted) for a
voltage dip down to 25% rated voltage at rated power.

In Fig. 2.36, a voltage dip down to 25% rated voltage for one second has
been recorded for a 6 MW turbine running at rated power. Also here, there are
only small differences between the new linear aerodynamic model and the model
based on a static c_P-λ table.

2.8 Summary

A simplified generic aerodynamic model for the use in system studies with converter based, variable speed wind turbines has been proposed. It is based on a linearized aerodynamic model.

The parameters required for the aerodynamic model are directly deduced from the physical response of turbine power to blade pitch angle change and rotor speed change. This allows a better understanding of the turbines response than an aerodynamic model based on c_P-λ tables where the same information is not visible directly.

The simulations show that the aerodynamic torque does not stay constant both following grid faults and following changes of wind speed. Aerodynamic models that are based on the assumption of a constant aerodynamic power may lead to unintended errors in dynamic studies.

The proposed model has been compared to aerodynamic models based on c_P-λ tables and a modified functional representation. There are only limited differences if the proposed model is compared to more detailed models with a larger number of parameters.

A comparison to measurements shows that the difference between the proposed and more detailed models is within the range of the measurement error. More detailed aerodynamic models are therefore usually not necessary for dynamic grid simulation studies.

3 Model of the Turbine Structural Dynamics

3.1 Introduction

This part gives a short overview of common representations of the mechanical system of wind turbines in simulation systems. Higher order representations of the turbine are compared to two-mass and single mass representations of the drive train and the relevance of the representation for the analysis of system stability studies is evaluated.

The most detailed level of simulation models of the mechanical structure are based on finite element models (FEM) of the entire turbine. These models are generally used for turbine design only. A reduction of model complexity is possible if only the dominant eigenfrequencies[4] of the turbine are considered. This is an approach used by the so-called multi-body simulation that defines the characteristics of different parts of the turbine (blades, tower, and drive train). Usually the first (lowest) two eigenfrequencies only of each of these components in each geometrical direction are considered in commercial turbine design software [35], [36]. Still, such models are not suitable for dynamic studies of the electrical system because they are too complex and too difficult to handle. These models still consist of several hundred up to several thousands of parameters.

The question analyzed in this section is which level of detail of the mechanical system is necessary for the use in dynamic electrical system studies. That means, the impact of the mechanical model on active and possibly reactive power output of the model during grid events needs to be regarded:

- In the first step, possible effects that could be modeled using detailed models are analyzed.

- In a second step, a single mass and a modified two mass drive train model are presented and compared.

3.2 Higher Order Representations of Drive Train, Blades and Tower

The wind moves the blades, creating a torque that is transferred to the generator through the drive train. Changes of torque can lead to changes of rotor speed and

---

[4] Generally, the word 'eigenfrequency' in this section refers to the equivalent eigenfrequency of a reduced order linear model. It is a simplified represents of the dominant coupled modes of the turbine that in reality always consist of an interaction of several components.

possibly to change of power output. As shown later in section 4.5, common approaches to active power control of wind turbines are torque control or power control including a drive train damper that modifies the power reference as function of measured generator speed. As a result, changes in rotor speed can lead to small changes in power output.

In Fig. 3.1 an overview of a wind turbine nacelle is given that shows the key elements of a wind turbine drive train. From left to right, hub, shaft, gear box, high speed shaft brake and generator are shown.

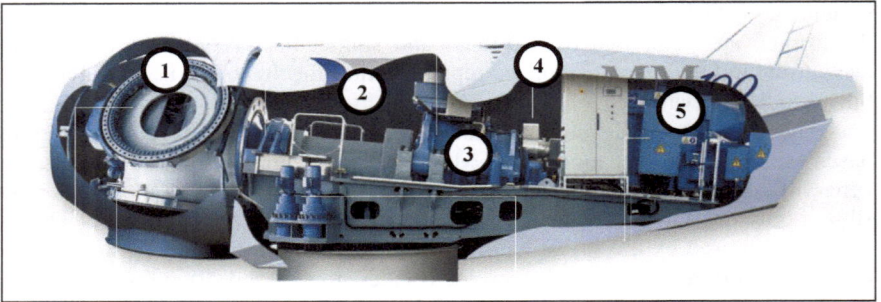

Fig. 3.1   Overview of turbine nacelle [37] with drive train consisting of (1) hub, (2) shaft, (3) gear box, (4) disk brake and (5) generator.

The following aspects of higher order representations are analyzed

- the gearbox

- higher order drive train eigenfrequencies

- detailed blade structural model

- tower shadow, rotor imbalance and rotor eigenfrequencies

- turbine component production tolerances and aging effects

- the tower representation

with respect to their impact on the power output, either by adding additional effects or oscillations or by modifying the expected turbine response.

3.2.1   Gearbox Representation

With an increase in length of the rotor blades, the speed of the rotor has to decrease because the speed of the blade tips should not be faster than about 70 m/s .. 80 m/s in order to limit the acoustic impact of the blades. The generator design by contrast requires sufficient speed for the induction. As a result most large wind turbines use gearboxes to increase the speed of the generator with respect to

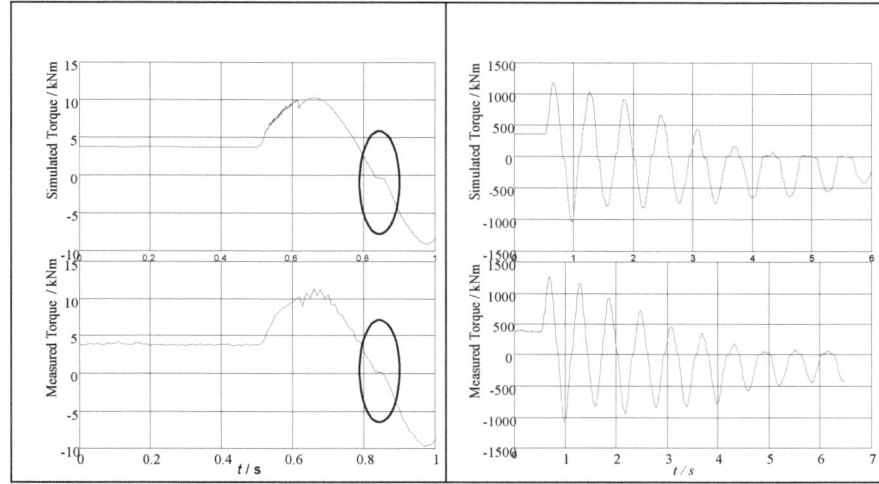

Fig. 3.2   Measurement and simulation of drive train high speed shaft torque of 1.5 MW wind turbine during forced shutdown[38].

Fig. 3.3   Measurement and simulation of drive train low speed shaft torque of 1.5 MW wind turbine during forced shutdown [38].

the turbine main shaft. If no gearbox is used, a large rotor diameter also requires a larger generator diameter and a higher number of generator pole pairs.

The gearbox is connected with elastic couplings to both the generator and to the rest of the nacelle. These couplings have a nonlinear stiffness that causes the eigenfrequency to change depending on the load condition.

A further nonlinear effect to be considered is the variation of the drive train eigenfrequency as function of the load. Following a change of the sign of the torque, the stiffness is undefined for a short period of time until the other sides of the gears get in contact again (see Fig. 3.2 and Fig. 3.3). Therefore, the resulting eigenfrequency of the drive train changes considerably if the sign of the torque changes.

These effects modify the first drive train eigenfrequency as function of the load. They could be modeled using detailed simulation models. But the input data is difficult to obtain since the gearbox usually consists of several stages. This means that the effect described above is actually a cascade of events of the different gearbox stages. The effect should to be taken into consideration when the eigenfrequency of the wind turbine is analyzed.

3.2.2   Representation of Higher Order Drive Train Eigenfrequencies

The first drive train eigenfrequency commonly describes the movement of a larger inertia (consisting of blades and hub) against a smaller inertia (consisting of generator and disk brake). Shaft and gear box have an impact on the stiffness, but have a negligible inertia compared to the rotor.

The second drive train eigenfrequency (sometimes referred to as first high speed shaft eigenfrequency) is used in some detailed simulation models. It represents the movement of the disk brake and the high speed stage of the gearbox against the generator on one side and turbine rotor on the other side. This eigenfrequency is generally > 5 Hz and small compared to the first drive train eigenfrequency. Under normal conditions it will not be relevant in the power output of variable speed wind turbines.

3.2.3 Detailed Blade Structural Representation

For the use in detailed simulation systems, generally only the first two eigenfrequencies of each blade in edgewise (chord wise) and flap wise mode are modeled

The most relevant eigenfrequency of the rotor, consisting of hub and rotor blades is the first in-plane torsional mode. It describes an in phase movement of the rotor blades (all three blades moving in phase to each other) in the first in plane eigenfrequency of the rotor blades against generator and the second lateral (sideways) tower eigenfrequency. (see Fig. 3.4 and Fig. 3.8). The oscillation of the tower results from a change of the generator torque that induces a change of the torsional momentum of the nacelle. The eigenfrequency can vary due to design parameters, production tolerances and depends on the stiffness of the foundation.

The stiffness of the blades and therefore the eigenfrequency in this mode (see Fig. 3.5) also depends on the blade pitch angle. Blades are far stiffer in edgewise mode than in flap wise mode. As a result, at 0 deg blade angle (almost only edgewise eigenfrequency contribution of the blades) the eigenfrequency of the rotor is higher than if the blades are pitched away from 0 deg (see Fig. 3.6).

As an approximation for the calculation for the first collective in plane mode, the in plane stiffness $K_{Blade\_in\_Plane}$ of a rotor blade could be described as function of blade pitch angle

$$K_{Blade\_in\_Plane} = K_{Blade\_Edgewise\_1} \cos(\Theta) + K_{Blade\_Flapwise\_1} \sin(\Theta) \qquad (3.1)$$

with $K_{Blade\_Edgewise\_1}$ and $K_{Blade\_Flapwise\_1}$ as the stiffness of the first edgewise and flapwise mode.

But as shown in Section 3.3.3, drive train oscillations of converter based wind turbines decay fast, and the impact of the changes of blade stiffness on the turbine eigenfrequency is limited, (see (3.8) in section 3.3.2) and does not play a role at most operating points. It is usually relevant only at very high wind speeds.

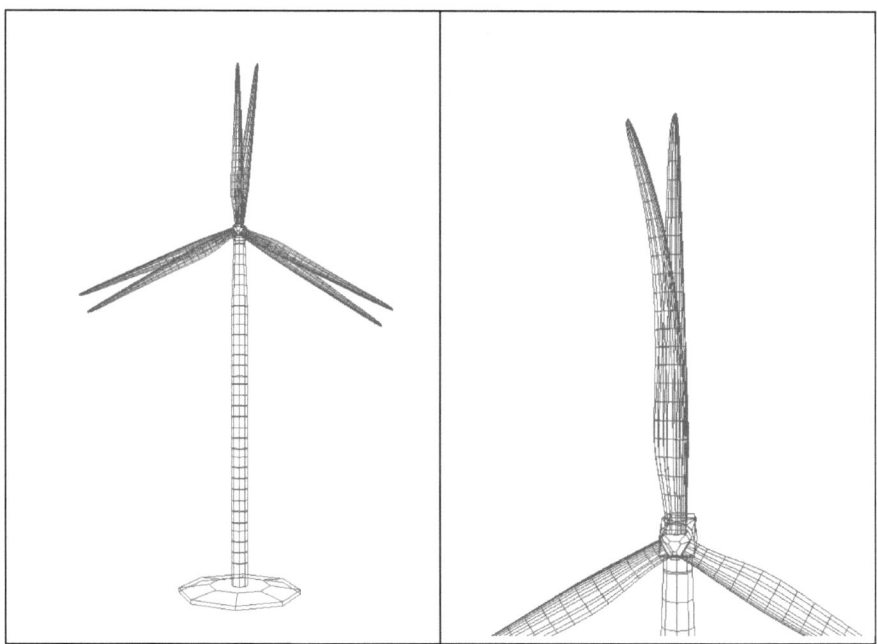

Fig. 3.4 First in-plane (torsional) mode of wind turbine shaft.

Fig. 3.5 First blade chord wise (edgewise) mode.

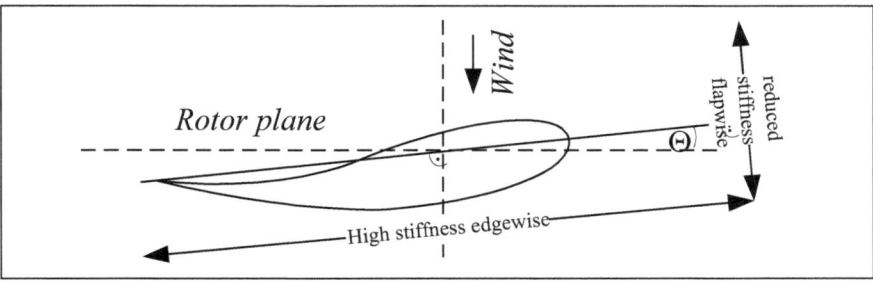

Fig. 3.6 Effect of blade angle changes on the in-plane stiffness of the rotor.

At high wind speeds, the pitch angle is > 0 deg; as result, the stiffness of the blades in the rotor plane will have decreased (see (3.1) and Fig. 3.6). As consequence also the first drive train eigenfrequency will decrease. If the effect of changes of the eigenfrequency of the blades at high wind speed is to be analyzed, a simulation with a slightly reduced drive train eigenfrequency is the most appropriate approach.

3.2.4 Tower Shadow, Rotor Imbalance and Rotor Eigenfrequencies

Some oscillations of power output are functions of the rotor speed. For example, the power reduction due to the wind speed reduction close to the tower ('tower shadow') affects the torque with 3 times the rotor frequency ('3p') for a wind turbine with three rotor blades. Rotor imbalances lead to events with single rotor frequency ('1p') dependency. The impact of the rotor eigenfrequency and tower shadow on the grid for fixed speed turbines has been discussed in several publications [10], [39]. Those effects can be measured and reproduced by simulations, but the corresponding energy content is very low compared to the key eigenfrequencies observable following grid faults. These effects can therefore be ignored for system stability studies.

3.2.5 Impact of Production Tolerances and Aging Effects

The rotor blades of a turbine need to have an almost identical weight distribution in order to avoid mechanical loads on the turbine as result of static or dynamic imbalance. In order to minimize production tolerances, blades are designed with chambers that can be filled (for example with lead) so that an even weight distribution along all the blades of a wind turbine can be achieved. Differences in the resulting weight and weight distribution of the rotor blades lead to differences in the eigenfrequency.

A second nonlinear effect that can be observed is the change of eigenfrequency and damping of the drive train as a result of the aging effects due to the operation time of the wind turbine. Both stiffness and damping especially of mechanical connections like couplings will change with age. Both effects are difficult to model and can currently not be predicted with reasonable effort. It is therefore necessary to take into account tolerances for the eigenfrequencies of the drive train.

3.2.6 Tower Representation

Grid faults with voltage drops lead to changes of rotor speed and of the electrical power output during and following the fault for wind turbines without energy recovery[5] during the fault. The increase of rotor speed and the resulting change in pitch blade angle modifies the thrust on the turbine and therefore has an impact on the first fore-after eigenfrequency of the tower [40], see Fig. 3.7. The induced fore-after tower movement influences the effective wind speed of the rotor.

These effects can be measured and reproduced using detailed simulation systems for the structural dynamics. For fixed speed wind turbines, the effect of

---

[5] Like DC-link choppers or passive devices of FSC wind turbines that are able to absorb the surplus energy from the DC-link that cannot be fed into the grid during a grid fault any more. For more details see section 5.4.4.

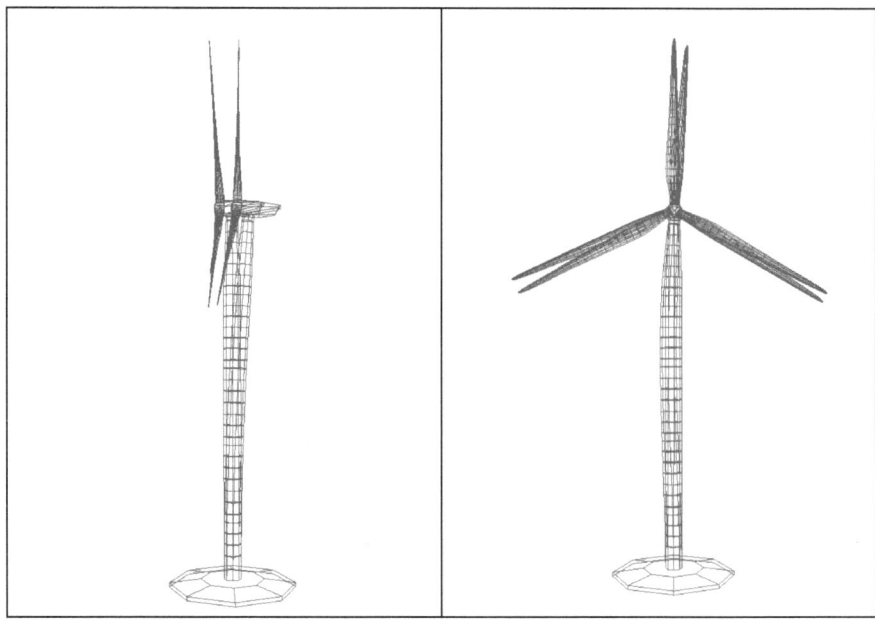

Fig. 3.7   First longitudinal (fore-after)   **Fig. 3.8**   Second wind turbine tower lateral
eigenfrequency of the wind turbine   (sideways) eigenfrequency.
tower.

the tower representation on flicker has been addressed in [41], its relevance in stability studies has been discussed in [42].

This effect of tower movements on the effective wind speed of variable speed wind turbines needs to be analyzed. The tower movement changes the effective wind speed and as result the rotor speed of the turbine. There is a feedback of rotor speed changes on the power output if the turbine control is based on a torque reference (see Section 4.5). As a result, the power output which is torque times rotor speed, varies with the oscillation of the tower following a grid fault. The frequency analysis of a measurement following of a voltage dip down to 0.0 p.u. voltage is shown in Fig. 3.9.

The first tower eigenfrequency is commonly in the range of 0.4 Hz .. 0.6 Hz for towers between 60 m and 100 m. The wind speed has a local maximum at 0.45 Hz - which represents the first tower eigenfrequency with an amplitude of 0.03 p.u.. The corresponding response of the power output is 0.002 p.u.. The response of the power is probably less than this value, the error is a result of the frequency analysis that does not filter out other frequencies in proximity of the target frequency. The oscillation is well damped with 0.25 p.u. decrease over one period (a logarithmic decrement of 0.29).The magnitude of the measured power

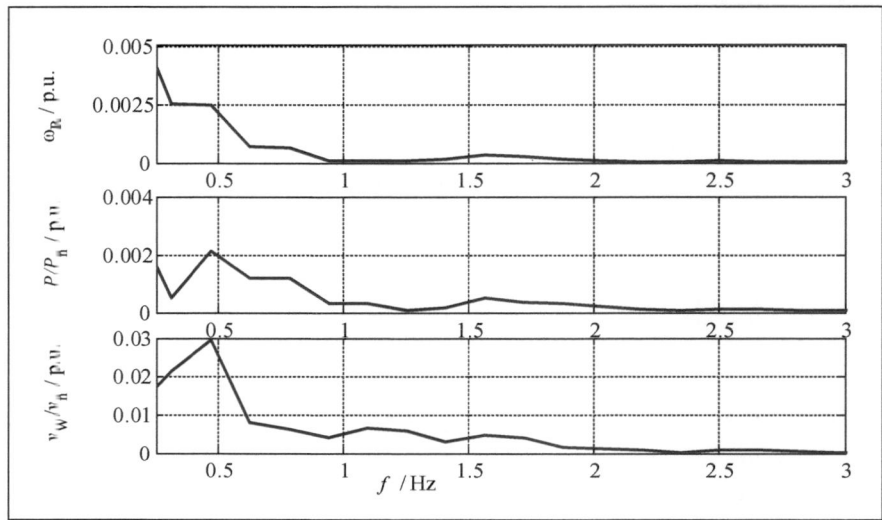

Fig. 3.9   Response of rotor speed, active power and wind speed to tower oscillations following a grid fault.

charge (0.002 p.u.) is very small compared to the power change following a grid fault (1 pu).

The values measured are in agreement with the simulations, as expected the effect on the power output is very small. Due to the decoupled control of active and reactive power of converter based variable speed turbines, there is no relevant feedback of grid frequency and voltage phase angle changes on the power output (see chapter 5). As result, grid frequency changes will not trigger tower oscillations, there will not be any resonance resulting from grid frequency and grid voltage phase angle changes under normal grid conditions.

The response of DFG and FSC converters without energy recovery is comparable; the impact of FSC with DC-link energy recovery is even far lower since almost no tower oscillation will be triggered. The effect of tower oscillations on grid stability can therefore be neglected for converter based variable speed wind turbines for towers up to 100 m. There is no relevant contribution in the frequency range of 0.2 Hz .. 2 Hz.

New tower developments, especially the use of hybrid (concrete + steel) tower allows the design of towers beyond 120 m up to currently 160 m even for turbines in the range of 2 MW .. 3.5 MW. A possibly lower eigenfrequency of such towers could slightly increase the impact of the tower movement on the power output following grid faults. If no countermeasures are taken - for example by using an improved control - the impact of higher towers would have to be reevaluated in future for very tall wind turbines.

3.3 Single and Two Mass Representation of the Drive Train

A single mass representation of the turbine rotor (including blades, hub, shaft gearbox - if present - and generator) is common for generic variable speed wind turbine representations [1]. Two-mass representation of variable speed wind turbines have been presented by [10] and [43], on the other side the use of single mass models is recommended by the authors of [21] and [44].

3.3.1 Single Mass Representation of the Drive Train

The mathematical description of a single mass drive train is given by

$$\frac{d\omega_R}{dt} = \frac{1}{2H_T}\left(T_{Aero}^* - T_{El}^*\right) \tag{3.2}$$

with ω_R as the rotor speed and H_T as the turbine inertia constant. The single mass drive train can be modeled as shown in Fig. 3.10.

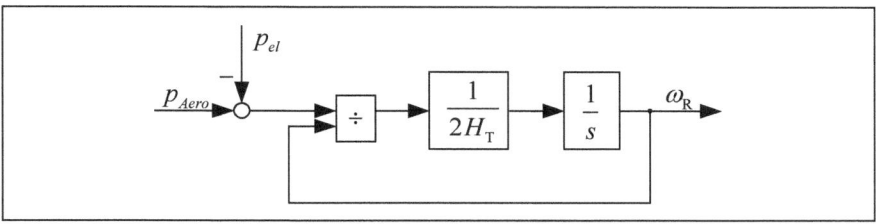

Fig. 3.10 Single-mass representation of the drive train.

The inertia time constant is defined as the kinetic energy stored in the drive train in Joule divided by the nominal power. The inertia time constant H for modern turbines is in the range of 3 s .. 5 s.

$$H_T = \frac{J_T}{2}\frac{\Omega_n^2}{P_n} \tag{3.3}$$

Note that the use of active power (W) and not apparent power (VA) is recommended for p.u.-scaling to derive normalized quantities. [6]

---

[6] This becomes relevant if the required reactive current capability during FRT is considered (see section 6). An additional passive compensation would increase the apparent rating of a wind plant. A specification based on apparent rating would then require additional active components to provide sufficient (up to 1 p.u.) reactive current during a fault.

3.3.2   Two Mass Representation of the Drive Train

For variable speed wind turbines, the drive train can generally be reduced to a two mass representation, with one lager inertia representing rotor blades and hub (described by the suffix WR for 'wind rotor') and one smaller inertia representing generator and disk brake (described by the suffix R for 'rotor'), see also [14], p.488. The key elements of a two mass drive train representation are shown in Fig. 3.11.

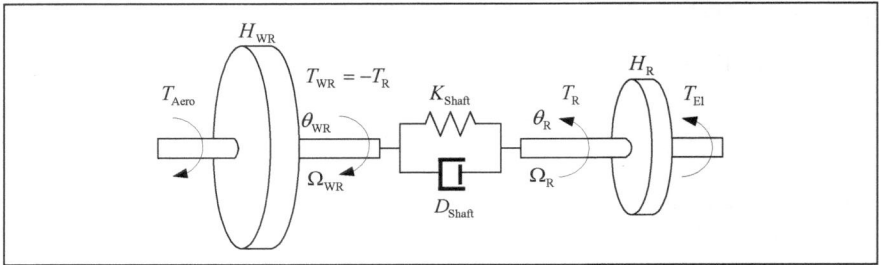

Fig. 3.11 Two-mass spring-damper representation of the drive train.

The equations of the two mass spring damper representation can be given as

$$\frac{d\omega_R}{dt} = \frac{1}{2H_R^*}\left(T_{Aero}^* - d_{Shaft}\left(\omega_{WR} - \omega_R\right) - k_{Shaft}\delta_{WG}\right) \tag{3.4}$$

$$\frac{d\omega_{WR}}{dt} = \frac{1}{2H_{WR}^*}\left(-T_{El}^* + d_{Shaft}\left(\omega_{WR} - \omega_R\right) + k_{Shaft}\delta_{WG}\right) \tag{3.5}$$

$$\frac{d\delta_{WG}}{dt} = \omega_{WR} - \omega_R \tag{3.6}$$

with all units in p.u.. The aerodynamic torque is calculated as $T_{Aero}^* = p_{Aero}/\omega_{WR}$, the electrical torque as $T_{El}^* = p_{El}/\omega_R$.

As shown before, the turbine rotor is not a stiff mass. Using a multi body system modeling approach, a generalized description of the drive train modes can be calculated using

$$[\mathbf{M}]\ddot{x} + [\mathbf{D}]\dot{x} + [\mathbf{K}]x = F \tag{3.7}$$

with [M], [D] and [K] as matrices describing inertia, damping and stiffness respectively.

A simpler approach using modified parameters of a two mass model for representing the impact of the blade eigenfrequencies on the resulting eigenfrequency is described in [10] that can be applied if key turbine parameters are known from a datasheet. The first drive train eigenfrequency f_{Shaft} is given as :

$$f_{Shaft} = \frac{1}{2\pi} \sqrt{\frac{\frac{1}{J_{WR}} + \frac{1}{\eta^2 J_R}}{\frac{1}{\left(2\pi f_{Edge}\right)^2 J_{WR}} + \frac{1}{K_{DT}}}} \qquad (3.8)$$

with J_{WR} as rotational inertia of wind turbine rotor (mainly blades and hub) J_R as rotational inertia of the generator rotor at the high speed shaft, η as gearbox ratio, f_{Edge} as the 1st edgewise eigenfrequency of the blades and K_{DT} as the stiffness of the drive rain (assuming a stiff rotor). The resulting modified stiffness K_{Shaft} is

$$K_{Shaft} = \Omega_{0Shaft}{}^2 J_{WT} = \left(2\pi f_{Shaft}\right)^2 \frac{J_{WR} \eta^2 J_R}{\left(J_{WR} + \eta^2 J_R\right)} \qquad (3.9)$$

The effect of damper windings (of synchronous generators) is part of the model structure of some generic models. However, it is not relevant for the representation of converter based variable speed wind turbines.

The damping coefficient D_{Shaft} is difficult to calculate analytically due to nonlinearities of the components of the drive train. It is recommended to calculate this value by evaluating measurements or the results of detailed simulations. It can be calculated by using the amplitude of the subsequent periods of a signal and calculating the dimensionless logarithmic decrement, assuming that one dominant eigenfrequency exists:

$$\delta_s = \ln\left(\frac{a(t)}{a(t + t_p)}\right) \qquad (3.10)$$

Using the damping ratio ξ defined as

$$\xi = \frac{\delta_s}{\sqrt{\delta_s^2 + 4\pi^2}} \qquad (3.11)$$

the damping coefficient is calculated as:

$$D_{Shaft} = 2\xi \sqrt{k_{Shaft} J_{WR}} \qquad (3.12)$$

Note that the eigenfrequency observed in simulations and measurements may differ from the frequency calculated in (3.8) due to the influence of the drive train damper (see section 4.5.2) and the nonlinear stiffness property of the power speed characteristic of the turbine control.

3.3.2 Comparison of Single and Two Mass Representations

Control settings

In Fig. 3.12 the generator speed for a single-mass and a two-mass model are shown. The difference in this case is about 0.04 p.u.. The maximum speed deviation between one- and two-mass models can be calculated if the worst case conditions are known (usually rated rotor speed, rated power and a voltage dip down to zero power). This means that using a one-mass drive train representation, a more conservative overspeed protection setting is required in the model.

The higher speed excursion of the two-mass representation may also lead to an increased pitch activity compared to the single mass representation. Therefore, an adaption of the settings of the pitch control parameters may be needed for a single mass drive train representation.

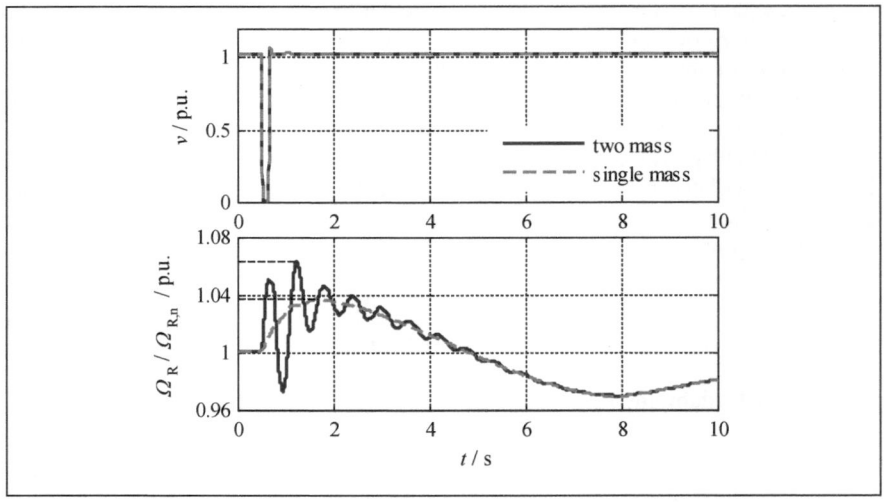

Fig. 3.12 Speed excursion of generator during voltage dip for single-mass (dashed) and two-mass models (solid).

Power output following grid faults

A comparison of voltage dip simulations using single mass and two-mass drive train models is shown in Fig. 3.13 for a 150 ms voltage drop to 0.0 p.u. and in Fig. 3.14 for a 500 ms voltage drop to 0.2 p.u.. In both cases the same turbine is modeled with the same parameters.

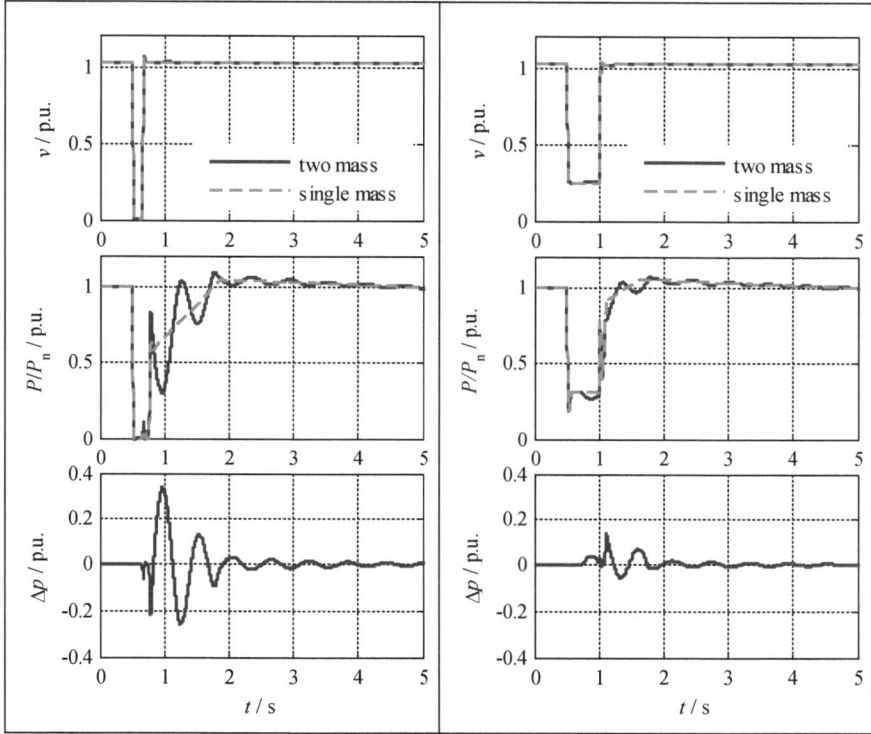

Fig. 3.13 Simulation of power fluctuations following a grid fault for single mass (dashed) and two-mass drive (solid) train representations for a voltage drop of 150 ms to 0.0 p.u. voltage.

Fig. 3.14 Simulation of power fluctuations following a grid fault for single mass (dashed) and two-mass (solid) drive train representations for a voltage drop of 500 ms to 0.2 p.u. voltage.

It can be seen that the duration of the fault has an influence on the oscillation after voltage recovery. The power oscillations following voltage recovery of the two-mass model decay within less than 3 s for both simulations.

The impact of an uncertainty of the drive train stiffness is shown in Fig. 3.15 and Fig. 3.16 for a voltage drop to 0.0 p.u. for 150 ms and 0.21 p.u. for 500 ms respectively. The resulting eigenfrequencies of the drive train are 1.65 Hz, 1.75 Hz and 1.85 Hz. The comparison shows that there is not only a phase difference of the post fault oscillations of power and speed, there can also be a difference in the amplitude. The reason is that depending on the moment of power recovery, the oscillation of the drive train may either be damped or accelerated. As a result of the differences in rotor speed, there is also a difference in output power.

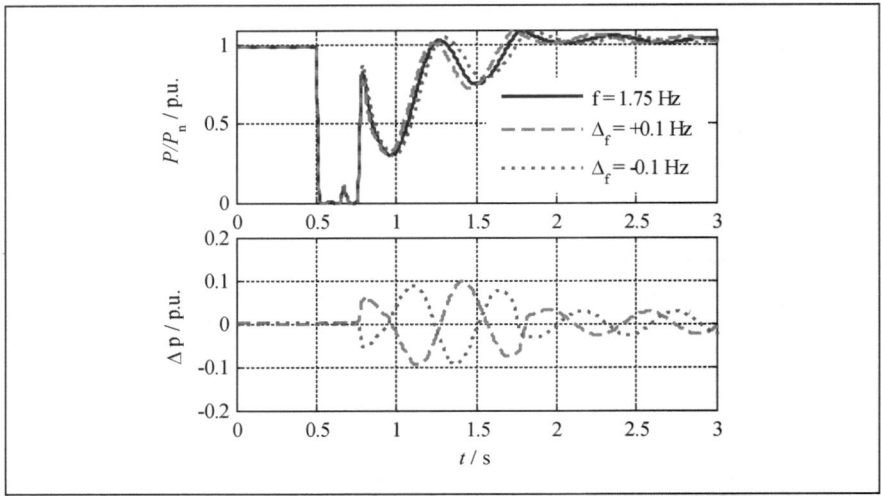

Fig. 3.15 Effect of different drive train stiffness (±5%) on post-fault power changes for a voltage drop of 150 ms to 0.0 p.u. voltage. The solid trace describes the response for an eigenfrequency of 1.75 Hz. An increase of the eigenfrequency by 0.1 Hz is described by the dashed trace, a decrease by 0.1 Hz by the doted trace.

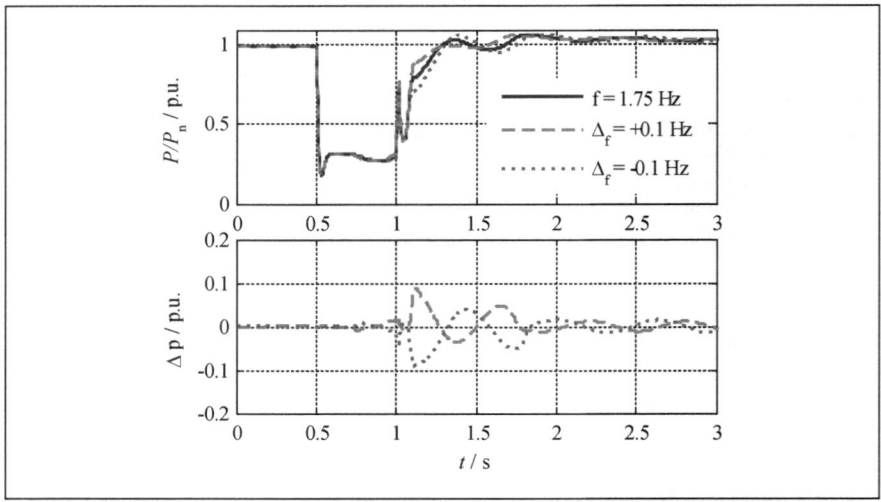

Fig. 3.16 Effect of different drive train stiffness (±5%) on post-fault power changes for a voltage drop of 500 ms to 0.2 p.u. voltage. The solid trace describes the response for an eigenfrequency of 1.75 Hz. An increase of the eigenfrequency by 0.1 Hz is described by the dashed trace, a decrease by 0.1 Hz by the doted trace.

The amplitude of the oscillation depends on fault duration, drive train parameters and the degree of voltage depression during the fault. (See also section 4.5.2 on active drive train damping).

If crowbar protection with small rotor resistances is used for DFG systems, the response to voltage dips may depend on the drive train representation. A single mass representation is stiffer than a more detailed model. As a result, the currents immediately following a voltage dip using a single-mass representation of the drive train will be higher than in reality. For such an analysis, also higher order generator models are required as described in [45]. These results can be relevant for the protection design, but the description of such currents is beyond the scope of stability studies.

3.4 Summary

A comparison between single mass and two-mass drive train representations for DFG wind turbines shows differences only during the first 3 s .. 5 s following a grid fault. A relevant difference exists mainly during the first two oscillations of the turbine drive train following voltage recovery (less than 1.5 s at common drive train eigenfrequencies in the range of 1.5 Hz .. 2.5 Hz). The oscillations of the rotor and resulting output power changes are usually well damped.

A single mass drive train model should usually be sufficient if frequency deviations and voltage stability are analyzed. In case a system is only weakly damped or has a low inertia and is responsive to power fluctuations in the range of 1 Hz .. 2.5 Hz, a two-mass model may be needed.

If a two-mass model is used, several simulations using a variation of the stiffness of the drive train are recommended. The reason is that both model uncertainties at varying wind speeds and blade angles as well as parameter uncertainties of the real turbines due to production tolerances (of blades, tower and foundation) and aging effects contribute to uncertainties in the drive train eigenfrequency.

Higher order turbine representations usually add only limited additional information at the expense of far more detailed models compared to a two-mass drive train. Representations using more than 2 masses may be necessary for fixed speed wind turbines or with directly coupled synchronous generators with variable speed gearboxes. A two-mass representation is accurate enough for representing all relevant effects for stability studies for DFG wind turbines.

4   Model of the Turbine Control

4.1 Introduction

The turbine control serves as the central interface to the pith system, the generator/converter system and to the plant controller as shown Fig. 4.1

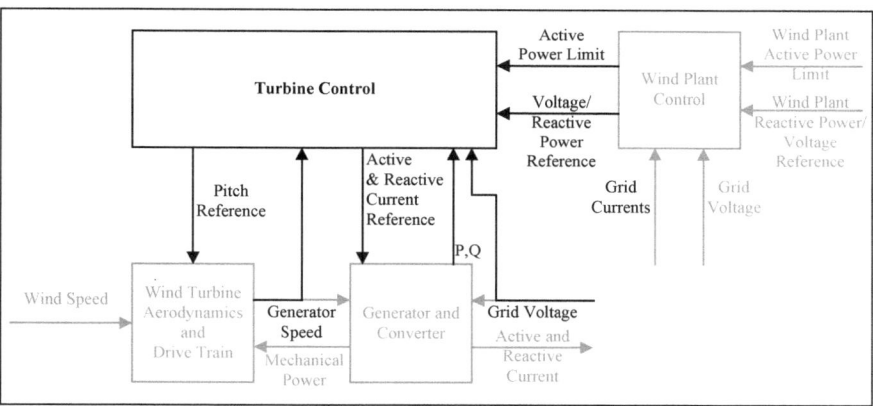

Fig. 4.1   Power control for fixed speed and limited speed range pitch controlled turbines.

It can be seen as a hierarchical system. On the highest level of control, we find the supervisory control, the safety system, and the communication system (see Fig. 4.2). The tasks of the supervisory is to provide reference values for the two main control loops, (1) the blade angle pitch control and (2) the converter torque control during the different operating modes, including (a) stand-by mode, (b) start-up mode, (c), power production mode and (d) shutdown mode

For the purpose of stability studies, only models of power production mode need to be considered. For power production mode, the main objectives of the variable speed active power controller design at turbine level are:

(1) power capture optimization

(2) reduction of extreme and fatigue loads on the turbine structure

(3) ride-through of grid faults

(4) power reduction on external request

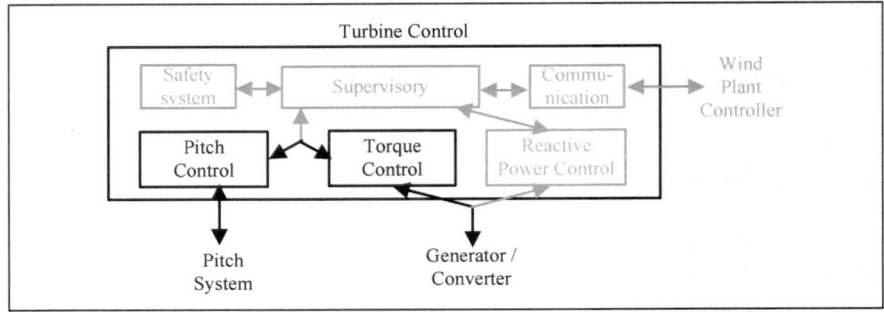

Fig. 4.2   Power control for fixed speed and limited speed range pitch controlled turbines.

Further demands based on new grid code requirements as (a) change of active power as function of frequency ('primary control'), (b) reserve or delta power control and (c) inertia-response to frequency changes are beyond the scope of this thesis.

This section is intended to

(1) give an overview of the relevant control approaches for fixed speed and variable speed wind turbines

(2) describe generic pitch and torque controllers that include concepts for handling ride-through of grid faults and power reduction on external requests and for decoupling the pitch-speed and torque speed control loops that both try to control speed

(3) show the difference between active damping control and a modeling through passive damping

Turbine reactive power control is discussed as part of the plant control in chapter 6.

4.2 Fixed Speed / Limited Speed Range Wind Turbine Control

The first generation of commercially successful wind turbines from the beginning of the 80 s had a generator directly coupled to the grid and was often using passive aerodynamic stall to limit the aerodynamic turbine power to rated power. As a system for emergency stop, besides the brake on the turbine shaft, a second independent braking system is needed. Therefore aerodynamic breaking was used; usually the blade tips were twisted to limit the aerodynamic power.

For turbines beyond 1 MW rated power, it is easier to control the blade angle directly at the hub instead of installing a system to change the blade angle at the tip only.

Active stall control moves the blade angle towards negative blade angles in order to improve the power limitation stall regulated turbines. To be able to stall, the mechanical torque of the drive train must be higher than the wind torque, otherwise the turbine rotor would accelerate. This is not a problem for directly coupled generators that are operated at fixed speed, because they can easily handle overload for several seconds.

Pitch control by contrast moves the blade angle in positive direction to limit the aerodynamic power. By allowing a limited speed variability of the rotor following a gust, a part of the additional energy provided by the wind can be converted to kinetic energy of the rotor. As a result, shaft and generator are exposed to a reduced torque increase only, thus limiting structural loads on the turbine.

A limited speed variability can be achieved using hydrodynamic couplings or by using a generator with a variable rotor resistance. Depending on turbine size and desired speed range, the resistors can be either included in the generator or connected via slip-ring to the generator.

Drawbacks of wind turbines with fixed speed or limited speed range are a reduced aerodynamic efficiency compared to other modern variable speed turbines due to the limited speed range and the need for addition components like STATCOMs to fulfill grid code requirements in many countries now.

Fixed speed or limited speed range wind turbines have no or limited capabilities to control the power output directly, the main input for the pitch control is the power output ([46], p. 50 and [14], p. 481). The power control structure for such turbines is shown in Fig. 4.3. An example of a pitch controller including

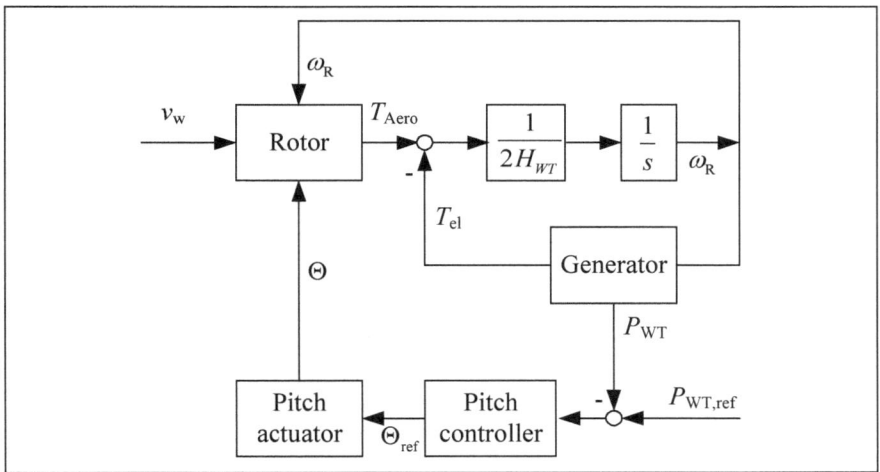

Fig. 4.3 Power control for fixed speed and limited speed range pitch controlled turbines.

pitch drive is shown in Fig. 4.4. The integral part of the pitch controller is limited if the blade angle of the pitch drive reaches its upper or lower limit.

Variable speed systems can make better use of the wind potential at low wind speeds and are able to further reduce structural loads at higher wind speeds. But they require power electronics which are far more limited in their overloading capability. In contrast to stall control, pitch control - turning the blades in positive direction - allows a far smother control of active power, this reduces the cost of power electronics and can also limit the mechanical stress on the turbine structure.

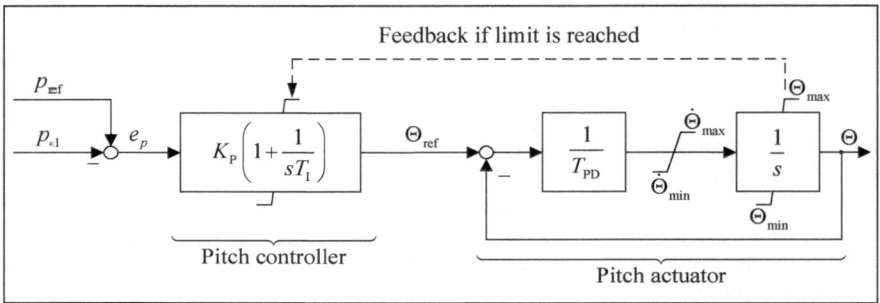

Fig. 4.4   Pitch control for fixed speed and limited speed range pitch controlled turbines.

This economic advantage of variable speed wind turbines had as consequence that both fixed speed systems and stall or active stall control do not play a role any more in modern converter based turbines beyond 1 MW rated power. There is extensive literature on the dynamics of fixed speed turbines for the analysis of existing energy systems ([28], [29], [39], [40], [41] and [47]).

4.3 Variable Speed Wind Turbine Control

4.3.1   Operation below Rated Wind Speed

The first objective for the design of a turbine controller is the optimizations of energy capture. Rotor blades that allow an operation at the optimum c_P-value throughout the speed range would be ideal. This would require a tip-speed-ratio of 5 deg .. 6 deg for common MW-size wind turbines (see Fig. 2.4). But design limitations [14] lead to tip-speed ratios generally no lower than 8-9 for modern turbines. The optimum steady state trajectory is shown in Fig. 4.5 (dashed curve through B and C). It describes an operation at optimum tip-speed-ratio λ_{opt} (see also Fig. 2.3).

Turbine structural limitations and the acoustic impact generally limit the maximum tip speed of rotor blades to around 70 m/s .. 80 m/s. Therefore, the

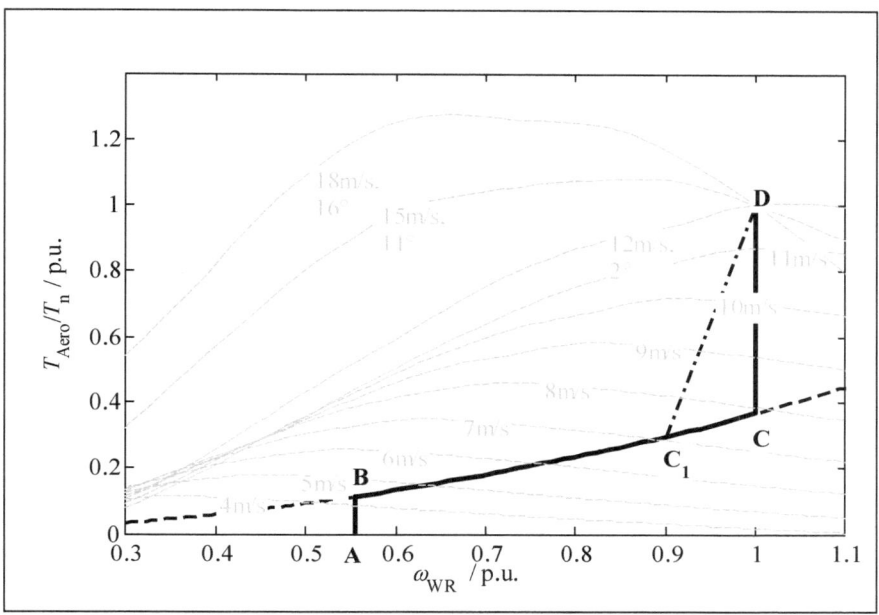

Fig. 4.5   Torque-speed curve of a variable-speed pitch-regulated turbine.

maximum rotor speed decreases with increasing rotor diameter. A lower speed
limit is given by the turbine losses, below 3 m/s .. 3.5 m/s the energy of the wind
cannot cover the electrical and mechanical losses of the turbine any more. The
lower speed limit may also be modified to avoid tower eigenfrequencies or as result of speed limitations given by the design of the generator and converter system of DFG turbines [48]. The solid line in Fig. 4.5 describes a typical steady
state trajectory for a wind turbine with limited upper and lower rotor speed range.

From a control point of view, each point on the trajectory is defined by wind
speed, rotor speed and power. As the wind speed fluctuates, different tracking algorithms are used to optimize the power tracking. Wind speed and wind direction
change as function of time and space, therefore the power captured by the rotor is
the integral over the entire rotor. Turbine wind measurement is usually on the nacelle using cup or ultrasound anemometers. But the area covered by measurement
does not even represent 0.01% of the rotor area. The measurement is also usually
behind the rotor and close to the nacelle, so additional distortions of the wind
signal have to be taken into account. As a result, wind speed may be an additional control input, but not the key controller input.

Control specifications could be based on torque or power as reference for
the converter. Torque control is more common since it offers an inherent damping of the drive train. Power control alone would lead to an unstable drive train

and needs to be combined with a sufficiently high active damping of the drive train (see also section 4.5.2)

Speed - Torque Control

One common control implementation is to specify the converter torque or power as function of rotor speed. For maximum power capture, the turbine needs to be operated at maximum c_P at $\lambda = \lambda_{opt}$ (see also Fig. 2.4 and Fig. 2.5). According to (2.14) rotor speed Ω_{WR} is then proportional to wind speed. Using (2.8) it can be shown that the power increases with Ω_{WR}^3, the torque increases with Ω_{WR}^2 :

$$T_{Aero} = \frac{P_{Aero}}{\Omega_{WR}} = \frac{1}{2}\rho A_{Aero} c_P v_W^2 \frac{R_{WR}}{\lambda} = \frac{1}{2}\rho \pi R_{WR}^3 \frac{c_P}{\lambda} v_W^2 \qquad (4.1)$$

With $v_W = \dfrac{\Omega_{WR} R_{WR}}{\lambda}$ the resulting control curve is

$$T_{ref} = \frac{1}{2}\rho \pi R_{WR}^5 \frac{c_P}{\lambda^3} \Omega_{WR}^2 - T_{loss} \qquad (4.2)$$

which is shown as line BC in Fig. 4.5. T_{loss} represents the mechanical and electrical losses. Close to rated speed the torque has to increase with a finite gain, so a slight loss in energy capture has to be accepted with this control strategy. This is represented by line C1-D in Fig. 4.5. The basic structure is shown in Fig. 4.6.

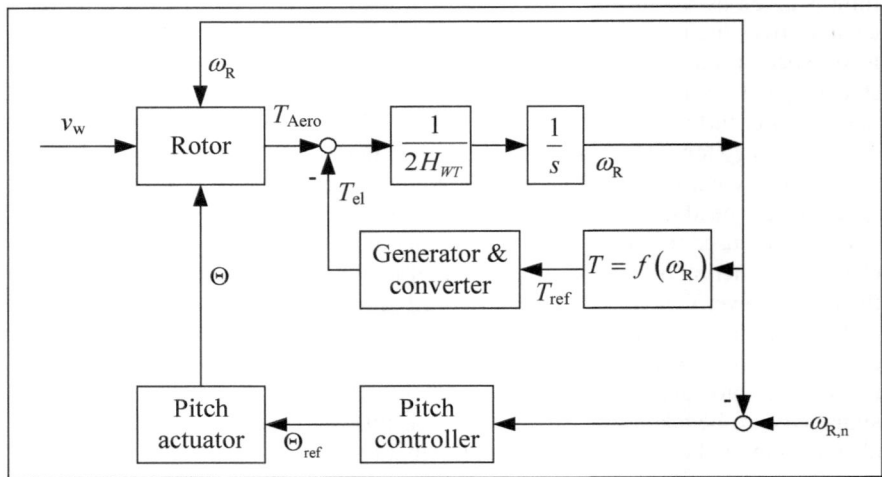

Fig. 4.6   Power control using speed-torque table.

Power - Speed Control

A different approach using PI-Control had first been proposed by [17]. Based on the measured turbine power, a speed reference is calculated using a lookup-table or analytical function. This speed reference is compared to the measured speed and its difference is used as input to a PI-Control that generates a torque reference (see Fig. 4.7). Operation between Points B-C1 remains the same, but an improved operation along lines C1-C and C-D is possible.

The control law for the power speed control can be derived by replacing torque with power in (4.2). The loss term T_{loss} which accounts for the losses in the drive train is represented by a reduced power factor c_P. This means the losses are approximated to be proportional to the power (see (4.3).

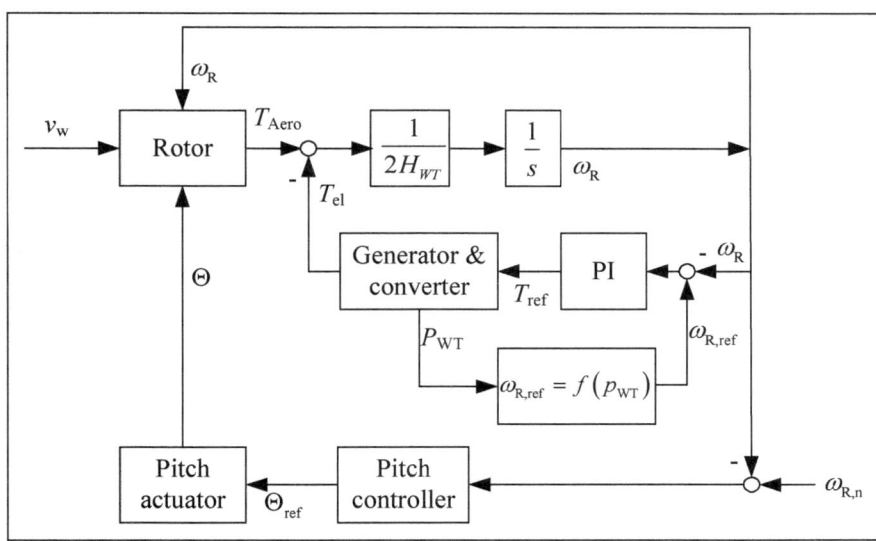

Fig. 4.7   Turbine power control using PI-control.

$$P_{WT,ref} = \frac{1}{2} \rho \pi R_{WR}^{\ 5} \frac{c_P}{\lambda^3} \Omega_{WR}^{\ 3} \qquad (4.3)$$

The control law with turbine power as input and speed reference as output is then:

$$\omega_{R,ref}(P) = \frac{\lambda}{\Omega_n} \left[\frac{2P_n}{\rho \pi R_{WR}^{\ 5} c_P} \right]^{1/3} P_{WT}^{\ 1/3} = k_{op} P_{WT}^{\ 1/3} \qquad (4.4)$$

The output of (4.4) needs to be limited to maximum and minimum stationary turbine speed.

The resulting rotor speed reference for a 2 MW wind turbine is shown in Fig. 4.8, the stationary relation of rotor speed as function of wind speed is shown in Fig. 2.10. Several strategies to optimize the maximum power tracking in unsteady wind conditions are discussed in literature ([14], p. 482ff), but are not relevant for the analysis of stability studies where constant wind speed is assumed.

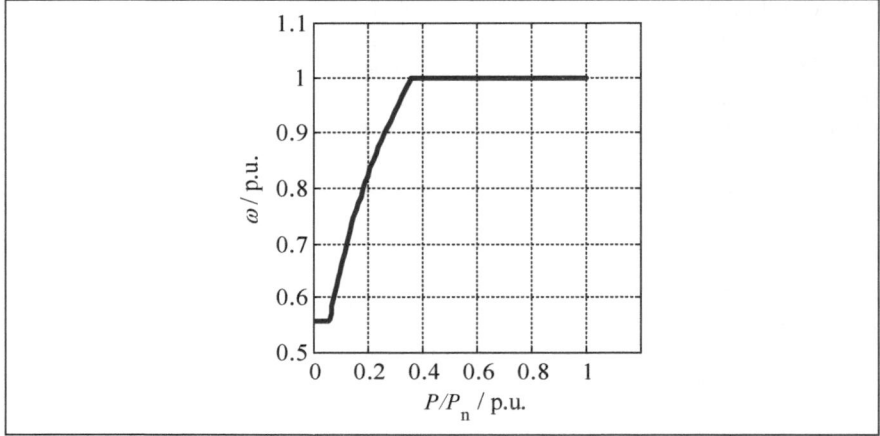

Fig. 4.8   Rotor speed reference as function of turbine power.

4.3.2   Operation at and Beyond Rated Wind Speed

Below rated wind speed keeping the blade pitch angle at optimum position of usually 0 deg is the only pitch activity required. Depending on the specific blade aerodynamics, a slight improvement of power capture may be possible by pitching towards slightly negative blade angles while operating between the points C-D at wind speeds below rated and slightly positive blade angles between A-B at very low wind speeds (see Fig. 2.4). But assuming a fixed blade angle of 0 deg between cut-in wind speed and rated wind speed will not lead to any relevant error for the system dynamics and is therefore common in wind turbine modeling. This allows the use of simpler aerodynamic models (see section 2) and reduces the number of parameters necessary for the description of the control system.

This simplifies the approach, below rated wind speed the turbine speed is only controlled by the converter. At rated wind speed, rated power of the turbine is reached, a further increase in wind speed needs pitch activity to limit the aerodynamic power to rated. This means that a change of control strategy is necessary beyond rated wind speed. The converter controls turbine speed below rated wind speed, above rated wind speed only the blade angle can control the rotor speed.

One of the key tasks in controller design of wind turbine is to find a good separation of these two control loops (torque-speed and blade angle - speed) in order to maximize power capture and limit turbine structural loads (see Fig. 4.9).

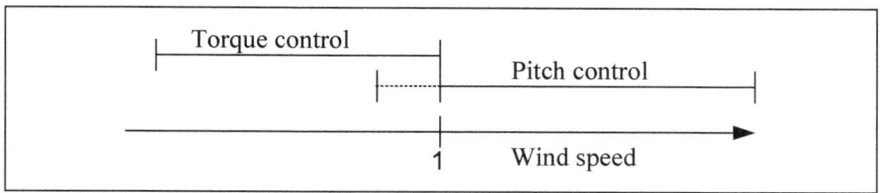

Fig. 4.9   Overlapping of the control ranges of torque and pitch control.

The rotor speed will fluctuate around rated speed as a result of gusts. This means the turbine speed will temporarily be below rated, which would cause the torque controller to reduce the torque reference. The consequence would be a reduced energy capture- an action that is clearly not desired at above rated wind conditions.

A simple approach is based on the definition of a minimum blade angle that blocks the torque controller from reducing the power reference. Assuming that a blade angle greater than zero means that the wind speed is still high enough to allow rated output, the torque reference would be kept constant until the blade angle reaches zero degrees again. On the other side, the pitch would be confined to zero degrees as long as the power is below rated.

More sophisticated approaches are optimized for operation with reduced external power reference and for grid faults. One possible approach is to have a cross-coupling between both controllers, keeping them active all the time but limiting their activity using additional coupling terms between the controllers ([14] p. 484).

A different approach proposed here is to add an offset to the reference values of pitch and torque controller. This ensures a realistic stable power production for all operating points.

4.4 Pitch Control Loop

The key task of the pitch controller is to limit the aerodynamic torque at above rated wind speeds and to keep the turbine speed in a limited operating range around rated wind speed. This is the task of the pitch-speed controller.

An additional pitch control loop is needed if an active power curtailment is demanded due to external requests from grid operator or from wind plant control. A response to frequency changes will usually be part of a faster control loop directly accessing the converter.

In order to limit the acceleration of the turbine during a grid fault, a fast control of the aerodynamic power following the voltage drop is necessary. A separate control loop is used to allow fast pitch activity to limit speed excursions.

4.4.1 Pitch-Speed Controller

The design of the pitch-speed controller is well documented in literature. A straight-forward approach is presented in [10], assuming operation at rated speed and constant power for the design. For a PI-controller in the form of

$$\Theta_{ref} = k_{PP}\left(1 + \frac{1}{sT_{IPP}}\right)\left(\omega_{R,ref} - \omega_R\right) \tag{4.5}$$

the parameters can be described as integral time constant:

$$T_{IPP} = \frac{2\xi}{\omega_0} \tag{4.6}$$

and proportional gain:

$$k_{PP} \cong \frac{2\xi k_{Shaft}\omega_{R,ref}}{\omega_0}\left[-\frac{dP}{d\theta}\right]^{-1} \tag{4.7}$$

with k_{Shaft} as the stiffness and the term $\left[-\dfrac{dP}{d\theta}\right]^{-1}$ as the inverse of the rate of change of power with respect to change of blade angle (see Fig. 2.13 and Fig. 2.17). This would require a gain scheduling of the controller in order to have a constant system gain for different wind speeds. For stability studies, a simplified approach using a fixed gain is acceptable since fast changes of wind speed are not considered. Common values are $\xi = 0.66$ and $\omega_0 = 0.6$ Hz [10].

Power Feedback to Decouple Control Loops

Both the pitch-speed control loop and the torque speed control loop try to control turbine speed. Since both control loops contain PI-controllers, a stable trajectory of operating points could be reached that do not provide the intended power output [49].

   An additional feedback is therefore used to ensure rated power is reached again for example following voltage dips. The turbine is operating at rated speed already before rated wind speed is reached (see Fig. 4.5 and Fig. 4.8). But once rated wind speed is reached, the pitch controller could become active. As a result, stable operating points with blade angles > 0 deg and active power < 1 p.u. could result. To prevent this, a feedback from the difference of active power and active power reference can be added to the input of the pitch controller that creates a negative offset als long as the final power reference value has not been reached. The modified control law is shown in (4.8)

$$\Theta_{\text{ref}} = k_{\text{PP}} \left(1 + \frac{1}{sT_{\text{IPP}}}\right)\left(\omega_{\text{R,ref}} - \omega_{\text{R}} - k_{\text{PX}}\left(p_{\text{max,ref}} - p_{\text{WT,ref}}\right)\right) \qquad (4.8)$$

The negative offset generated by $k_{\text{PX}}(p_{\text{max,ref}} - p_{\text{WT,ref}})$ is added to the pitch reference and forces the blade angle to lower values and in fact keeps it at zero degrees. Once the power reference (usually nominal power) has been reached this feedback has no relevance any more.

4.4.2 Pitch Compensator

External requests for power reduction as well as improved operation at lower wind speeds can be handled by the pitch compensator. Older turbine control designs using torque-speed control (see Fig. 4.5) could limit the output power by calculating an equivalent speed reference using an inverse of the torque-speed table. If power-speed control is applied, a different approach is needed. Using a PI-control that acts on the difference of power reference and measured power (or internal power reference, which is almost the same) allows a simple implementation of this control function.

$$\Theta_{\text{C}} = k_{\text{PC}} \left(1 + \frac{1}{sT_{\text{IPC}}}\right)\left(p_{\text{max,ref}} - p_{\text{WT,ref}}\right) \qquad (4.9)$$

The output Θ_{C} of (4.9) is limited to positive blade angles. In case fast power control is required, an additional direct control of the power reference may be used.

4.4.3 Pitch FRT Boost

Grid voltage drops lead to an immediate reduction of the maximum power the turbine can feed into the grid. A very short grid fault would not necessarily require a fast turbine response. Due to a turbine inertia of 4 s .. 5 s only a limited acceleration of the rotor would be expected for common faults of 150 ms or less. But it is not clear how long a fault is really going to last, therefore it is necessary to limit the aerodynamic power to avoid excessive acceleration.

Turbine controllers usually contain large numbers of nonlinear controls to optimize operation, for example to change the controller gain during faults. A simple alternative that allows keeping the pitch-speed controller untouched and still delivers acceptable results is to add a separate control loop that reacts proportional to the power reduction enforced by the voltage drop.

The input is the difference between demanded power and measured power. Using a simple p-control this can serve as a reference for the expected acceleration of the turbine. This control should stop once the fault has cleared. During very long faults, keeping the control loop active all the time could lead to very high blade angles that would lead to a sharp reduction in rotor speed and power following voltage recovery. Therefore a timer limits the activation time during long grid faults.

4.4.4 Pitch Actuator Model

The pitch actuator moves the pitch blades into the position requested by the turbine controller. Pitch actuators can be based on electric drives or on a hydraulic system. Commonly, a local pitch position controller receives reference values from the plant controller and ensures, that the desired blade position is reached [50] A pitch actuator model as shown in Fig. 4.10 needs to represent:

(1) communication and pitch position controller delays

(2) pitch speed limitations, generally with different pitch speeds in positive (limiting aerodynamic power) and negative (increasing aerodynamic power) direction. Pitch speed during turbine normal and emergency shutdown as well as turbine start are not represented by this model.

(3) blade pitch angle limitations, usually zero degrees, and a maximum blade angle at high wind speed. Negative blade angles that may be used to optimize operation during partial load as well as high positive blade angles during shutdown and turbine start are not represented by this model.

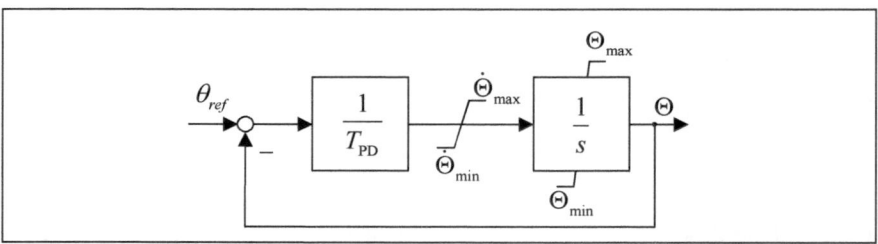

Fig. 4.10 Pitch actuator model.

4.4.5 Pitch Controller and Actuator Models

A model representation of pitch controller and pitch actuator is shown in Fig. 4.11. The signals for the pitch-speed control loop, the pitch compensator and the pitch FRT boost are added and supply the reference for the pitch actuator.

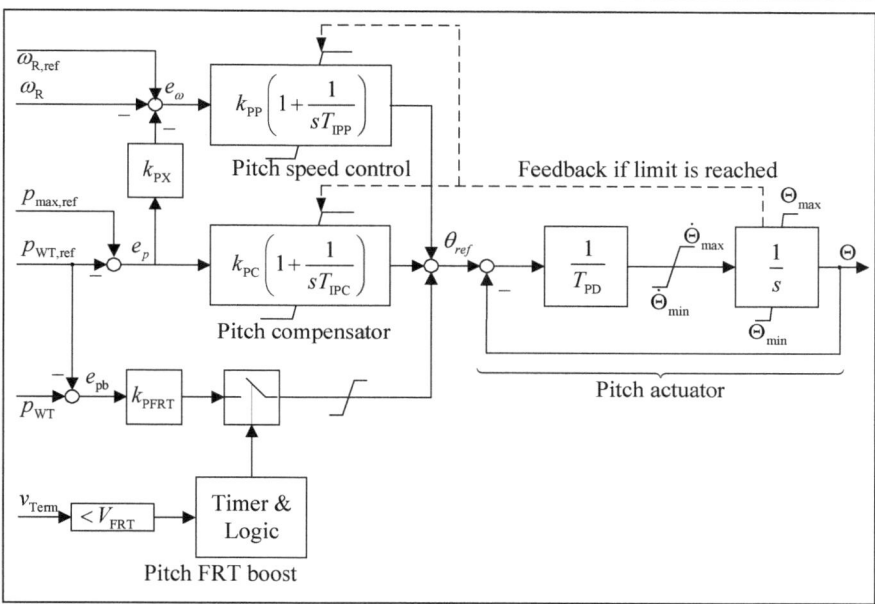

Fig. 4.11 Pitch controller and actuator model.

4.5 Torque/Active Power Control Loop

4.5.1 Power or Torque PI Control

The main task of the active power or torque control loop is to provide speed control below rated wind speed conditions. Above rated wind speed, the power or torque value shall be kept constant. Power control or torque PI control could be used, but generally torque control provides a more stable operation. In the case of simulating a flexible drive train, the increase of generator speed should be damped by an increase of torque.

Power Control

In the case of power control, a flexible drive train is not damped. Assuming a constant power reference p_0 and a slight increase $\Delta\omega_R$ of rotor speed, the torque reference t_0 is reduced in order the keep the power constant:

$$t' = t_0 - \frac{p_0}{\Delta\omega_R} \qquad (4.10)$$

As result of the torque reduction, the generator will now accelerate even more.

The opposite happens if the generator slows down a bit. In order to mini-mize oscillations, the torque should be reduced - instead it will be increased to keep the power constant. As a result, power control will always require either an additional damping controller (see following section) or a control function using a power-speed curve that is equivalent to a torque-speed curve, that means that power must increase at least in linear relation to speed.

Torque Control

Torque control does not damp directly, but it leads to a rather stable operation if the mechanical damping is not too low. In the case of grid faults, resetting the torque controller and initializing it with the torque available during the fault can help to avoid a windup of the integrator and limits overshooting of power follow-ing voltage recovery (see Fig. 4.12).

As shown above, a pitch compensator can be used to limit active power up-on external request. A different approach to limiting turbine power is to limit the output of the torque controller on external request. Since the aerodynamic power stays constant, the rotor speed and therefore the blade angle will increase.

Decoupling of Torque and Pitch Control

An equivalent to the decoupling of the pitch controller is also needed for the torque controller. As a possibility to decouple pitch and torque control, the speed reference of the torque controller may be limited to a value below rated speed for wind speeds above rated wind speed using an offset ω_{offset}. As a result, the torque controller will always be at its limit if the turbine operates close to rated speed. A short term reduction of rotor speed, for example due to gusts, will therefore not lead to an immediate reduction of the torque reference. This speed offset value may be calculated as (table) function of wind speed or blade pitch angle, the real implementation is manufacturer specific.

4.5.2   Effect of Drive Train Damping on Power Output

For variable speed wind turbines, the drive train can generally be reduced to a two mass representation, with one lager inertia representing the rotor blades and hub and on smaller inertia representing generator and disk brake. ([14], P.488). If two-mass-models are used, the damping of the drive train needs to be considered.

Variable speed wind turbines are often equipped with a drive train damping functionality based on a manufacturer-specific control. But if it is actually neces-sary to implement an active damping functionality in the model needs to be ana-lyzed.

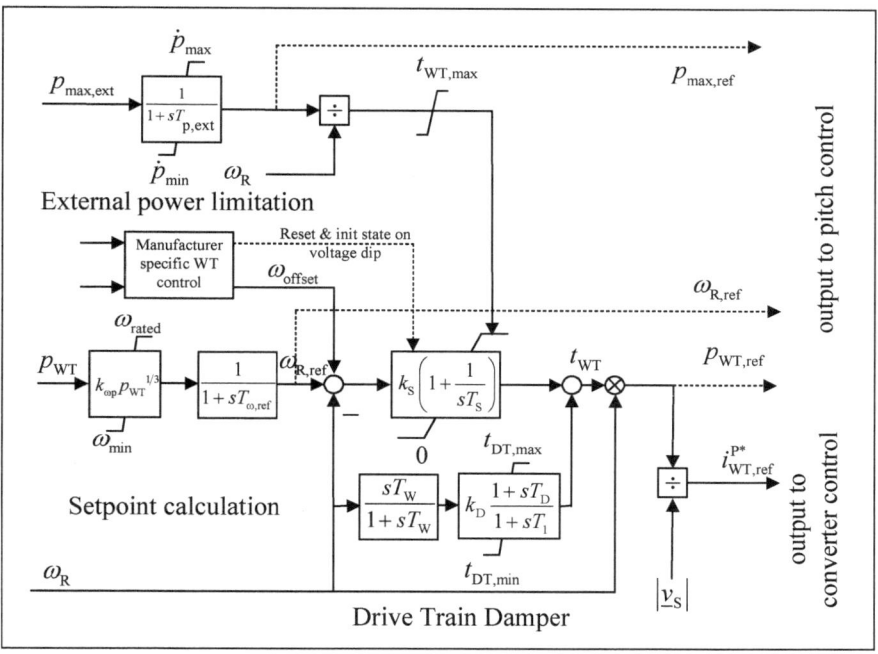

Fig. 4.12 Setpoint calculation and power control model including drive train damper and reference value calculation for the pitch control loop.

Active Damping Control of Drive Train Torsional Vibration

A common approach for active drive train damping in wind turbines is based on a second order function [14]

$$G(s) = k \frac{2\xi\omega s(1+s\tau)}{s^2 + 2\xi\omega s + \omega^2} \tag{4.11}$$

with ω as eigenfrequency, ξ as damping and τ as rewrapping that compensates for time delays of measurement and turbine controller. This approach takes into account that generally only the first shaft mode needs to be damped as it contributes the key energy. If a two-mass representation of the drive train is used, higher eigenfrequencies on the drive train will not play any role. A definition based on a washout filter (to describe the frequency to be damped) a gain and a lead-lag block (to define the phase angle correction) is commonly used in power systems and is applied here

$$G(s) = \frac{sT_W}{sT_W + 1} \cdot k_D \cdot \frac{1 + sT_D}{1 + sT_1} \tag{4.12}$$

Use of an equivalent passive Damping of Drive Train Torsional Oscillation in the Model

In [21] active damping is compared to simply increasing the mechanical damping factor of the drive train model. Key limitations of increasing the passive damping in the model to emulate the response of an active drive train damper are:

■ In case only faults with a fixed duration are considered, the result may be comparable to simulations with drive train damper or to measurements. But faults with a longer duration (with in reality possibly lower damping during the fault) show considerable differences.

■ Even though the damping effect on the drive train may be reproduced correctly, there is a difference in the power output. This is due to the fact that an active damper has an impact on the power output, but the impact of a passive damper is not visible directly in the power output.

Comparison of Model Implementations using Active Damping Control and passive Drive Train Damping

Fig. 4.13 and Fig. 4.14 show a comparison of two model implementations using (a) an active drive train damper based on (4.11) and (b) passive damping by increasing the damping coefficient of the drive train model on the power output following a grid fault.

The maximum error between active damping and passive damping is around 10% for the first oscillation period. After that, the maximum deviation is less than 3% - probably less than the expected accuracy of the model for dynamic changes. The main difference in power output between active and passive drive train damping is due to a phase shift in the power output. In most cases, an active drive train damper implementation is therefore not necessary.

In case the active power during fault recovery is of special interest, an active drive train damper implementation in the model will lead to an improved accuracy compared to real turbines especially for the first period of the drive train oscillation.

Generally, as a first step the relevance of the drive train eigenfrequency for the grid study should be estimated by varying the drive train eigenfrequency. If the eigenfrequency of the power output is critical, a generic drive train damper may not be sufficient. In this case manufacturer specific information will usually be necessary.

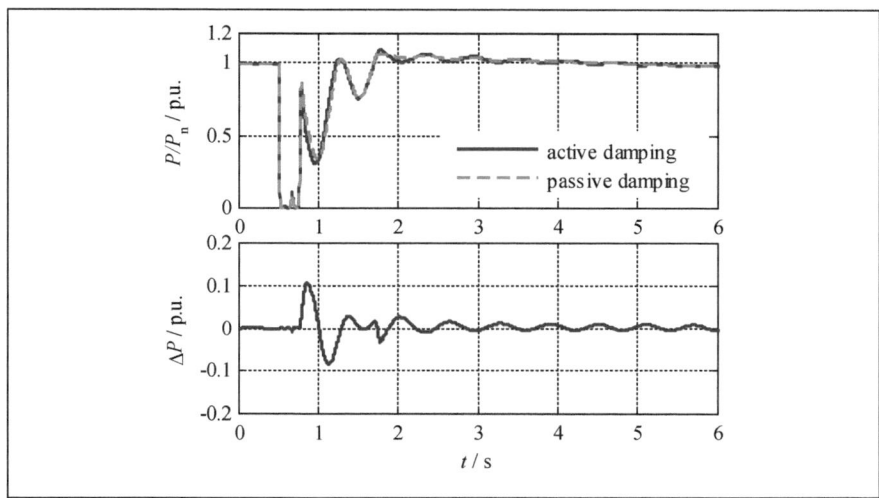

Fig. 4.13 Power recovery following 150 ms voltage dip to 0% of two mass model with active and passive damping. Active power and power difference are shown.

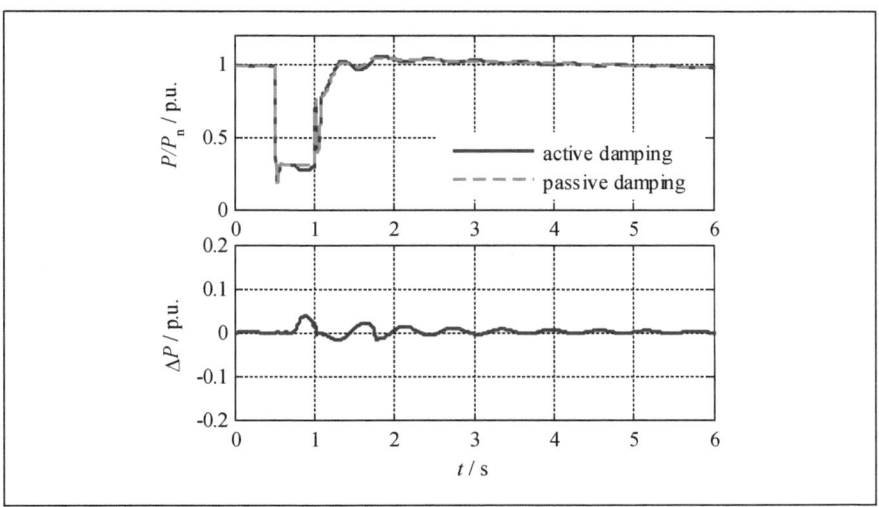

Fig. 4.14 Power recovery following 500 ms voltage dip to 21% of two mass model with active and passive damping. Active power and power difference are shown.

4.6 Summary

Generic control structures for wind turbine active power control are derived for the blade angle - speed and torque - speed control loops. Below rated wind speed, a maximum power tracking algorithm is employed by the torque controller. Above rated wind speed, the blade pitch controller limits the power output to rated power. Around rated wind speed, during power curtailment due to external reference and following voltage dips, care has to be taken to separate the activities of the pitch-speed and torque speed control loops. The impact of a cross-coupling of the torque - speed and blade angle - speed control loops is shown and is taken into consideration for both pitch and torque control design.

Turbine idling, start and shutdown are not relevant for system stability studies and therefore not covered by the proposed models. The models allow the parameterization of different, manufacturer specific control algorithms for power curtailment on external reference and for the power recovery following grid faults.

The impact of using an active drive train damper in the model is compared to a passive implementation in the model. By increasing the damping coefficient of a two-mass drive train model, the response of an active drive train damper can be emulated, and a comparable damping of the shaft speed can be achieved in the model. But if the active power oscillations following voltage recovery of a fault are of specific interest, an active damper implementation is recommended.

5 Generator and Converter

5.1 Introduction

Generic models for DFG and FSC based wind turbines have been proposed by WECC [4]. Simulation results of the DFG were compared to simulations with a more detailed model in [21]. Those models were so far mainly used in the USA, and the models were able to represent wind turbines with respect to common US grid code requirements. European grid code requirements often have specific requirements on reactive power delivery during and following grid faults [23], [51], [52]. New regulations in Germany, [53], [54] and Spain [55] require the validation of wind turbine simulation models with measurements of voltage dips. But this is beyond the scope of the existing generic models.

A very detailed approach to modeling DFG systems had been presented in [21], but the level of detail requires both a large number of parameters and integration step times smaller than 1 ms. A new approach is presented here that tries to come close to the accuracy of detailed models while keeping a simple structure and a very limited number of parameters as it is common for generic models. This is achieved on the basis of a detailed analysis of the generator equations and the converter control design.

The model can be used for large scale stability type RMS simulation and with an integration time step of 1 ms to 10 ms. Time constants of less than 5 ms have been eliminated by aggregating the generator and controller equations.

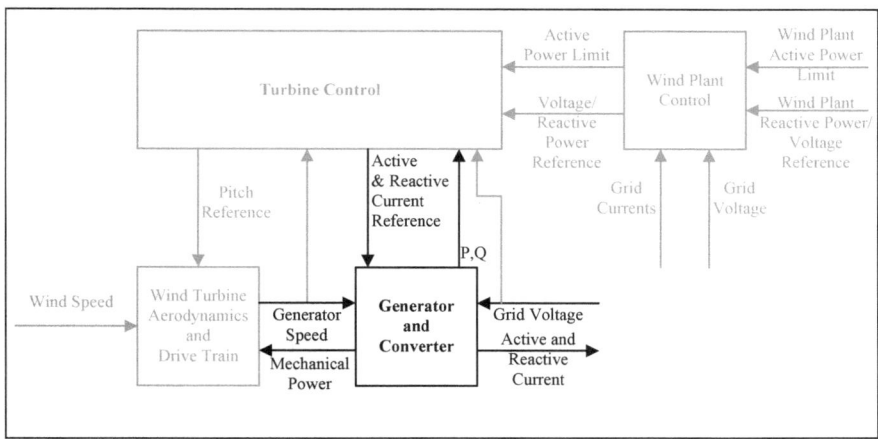

Fig. 5.1   Generator model as part of turbine model structure.

The generator and converter model as part of the wind turbine model is shown in Fig. 5.1.The model is intended to represent all aspects of a generator and converter that have relevance for large scale stability type RMS simulation. Some of the key issues it has to reproduce properly are:

■ Response to voltage amplitude and phase angle changes

■ Response to active and reactive power set point changes

According to the objective, transient effects should not be modeled in detail. During transient periods it is sufficient to represent fast changing variables by their average. The model is supposed to have a simple structure and should only use a limited number of parameters. On the other side, the physical background of parameters should remain visible, so that the value of parameters can be traced back to a physical or control quantity. The model should be simple enough not to disclose manufacturer specific know-how while still being accurate enough to allow a certification required by several grid codes.

The comparison of simulations and measurements of modern DFG and FSC based wind turbines shows that there are only limited differences between both technologies with respect to active and reactive currents during and following grid faults. Therefore, a modular approach has been chosen that allows the model to be used both for DFG and FSC model simulation using different parameters. The consumer oriented sign convention is applied for all equations and diagrams where consumed active power and inductive reactive power are considered to be positive.

5.2 Model of the DFG

The DFG is the most commonly used device for wind power generation. The stator is connected to the grid, the rotor terminals are fed in normal operation with a symmetrical three-phase voltage of variable frequency and amplitude. This voltage is provided by a voltage source converter usually equipped with IGBT based power electronics circuitry. The basic structure is shown in Fig. 5.2.

Limitations of DFG Designs without DC-Link Energy Absorber

In older DFG designs, protection against over-currents and undesirably high DC voltage was provided by the crow-bar (CR) placed on the rotor side of the machine side converter (MSC). Very high rotor current following voltage dips could cause the DC voltage to exceed the upper threshold allowed. The CR thyristor switches were then activated and the rotor terminals short-circuited through the CR resistance [29]. As a result, the DFG then is operated as a conventional slip-

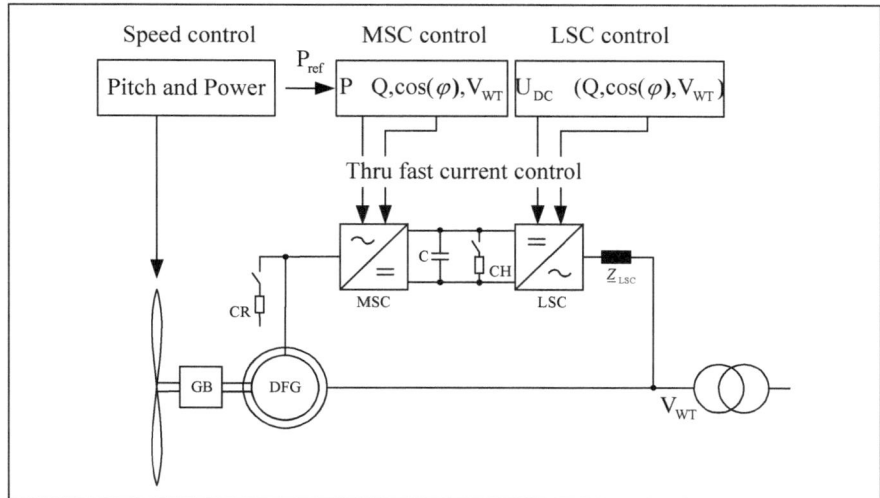

Fig. 5.2   Key components of DFG system.

ring induction machine without any control capability left for the MSC and the line side converter (LSC).

An improved design allowed for faster control of the DC - link voltage, so at least the LSC could remain in operation while the crowbar was active and compensate some of the reactive power requirements of the generator during the fault. IGBT-based crowbar designs of some manufacturers allows a shorter activation time of the crowbar, thereby reducing the time the MSC has to be deactivated [56].

The reaction to grid faults of all these designs is highly nonlinear and usually requires detailed EMT - models to exactly represent the DC-link that leads to the triggering of the crowbar. In [21] a modeling approach was presented that allows a detailed simulation even with RMS models. Still, a rather detailed model and very small simulation step times are needed.

Improved Protection of Modern DFG

New DFG designs use a higher rating of the MSC and passive or active DC-link elements. One common approach is using an IGBT switched resistor connected to the DC-link, referred to as "chopper" that limits the voltage in case the power fed into the DC-link is higher than the power the LSC can feed into the grid. As a result, an activation of the crowbar following grid faults as result of an increase of the DC-link voltage is no longer necessary. MSC and LSC remain controllable during the entire fault. IGBTs may block due to overcurrents, but this is only for very short periods that do not have relevant impact on the controllability of MSC and LSC. Crowbar-based converter protection may still be available as backup protection, but it is not activated during grid faults.

For the modeling approach selected, the DC-link voltage is therefore assumed to remain close to its nominal value and therefore does not need to be modeled explicitly.

5.2.1 Basic DFG Equations

The voltage and flux equations of the DFG in a dq coordinate system rotating synchronously with the grid frequency ω_0 are given by (5.1) - (5.4)

$$\underline{v}_S = r_S \underline{i}_S + \frac{d\underline{\psi}_S}{dt} + j\omega_0 \underline{\psi}_S \tag{5.1}$$

$$\underline{v}_R = r_R \underline{i}_R + \frac{d\underline{\psi}_R}{dt} + j(\omega_0 - \omega_R)\underline{\psi}_R \tag{5.2}$$

$$\underline{\psi}_S = l_S \underline{i}_S + l_h \underline{i}_R \tag{5.3}$$

$$\underline{\psi}_R = l_m \underline{i}_S + l_R \underline{i}_R \tag{5.4}$$

with $l_S = l_m + l_{\sigma S}$ and $l_R = l_m + l_{\sigma R}$ and the complex space vectors $\underline{\psi}_S = \psi_{Sd} + j\psi_{Sq}$, $\underline{\psi}_R = \psi_{Rd} + j\psi_{Rq}$, $\underline{v}_S = v_{Sd} + jv_{Sq}$, $\underline{v}_R = v_{Rd} + jv_{Rq}$, $\underline{i}_S = i_{Sd} + ji_{Sq}$, $\underline{i}_R = i_{Rd} + ji_{Rq}$.

In near steady state operation with both derivative terms $d\underline{\psi}_S/dt = 0$ and $d\underline{\psi}_R/dt = 0$ set to zero,

$$\underline{v}_S = (r_S + j\omega_0 l_{\sigma S})\underline{i}_S + j\omega_0 l_m (\underline{i}_S + \underline{i}_R) \tag{5.5}$$

$$\frac{\underline{v}_R}{s} = \left(\frac{r_R}{s} + j\omega_0 l_{\sigma R}\right)\underline{i}_R + j\omega_0 l_m (\underline{i}_S + \underline{i}_R) \tag{5.6}$$

$$s = \frac{\omega_0 - \omega_R}{\omega_0} \tag{5.7}$$

can be transferred into the well-known equivalent representation in Fig. 5.3:

5.2.2 DFG Model Representation

For stability type simulation, only the stator flux derivative is set to zero, $(d\underline{\psi}_S/dt = 0)$. Eq. (5.1) can be rewritten after some algebraic manipulation and

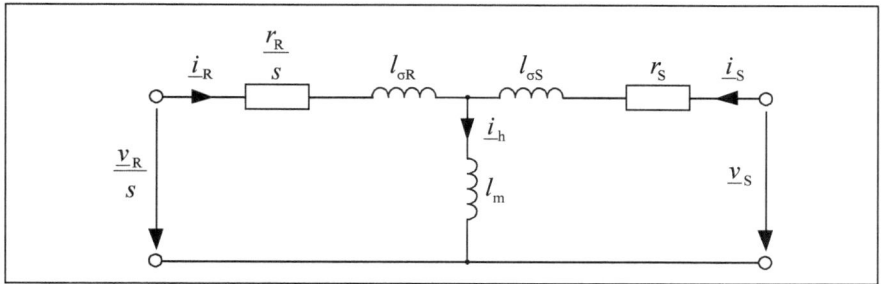

Fig. 5.3 DFG single line representation.

by using (5.3) and (5.4) for a Thévenin equivalent representation as shown in Fig. 5.4 to

$$\underline{v}_S = \underline{z}\,' \underline{i}_S + \underline{v}\,'$$ (5.8)

The internal transient impedance can be defined as

$$\underline{z}\,' = r_S + j\omega_0 l'$$ (5.9)

with

$$l' = l_S - \frac{l_m^2}{l_R}$$ (5.10)

and

$$\underline{v}\,' = j\omega_0 \frac{l_m}{l_R}\underline{\psi}_R = j\omega_0 k_R \underline{\psi}_R$$ (5.11)

as the corresponding transient driving Thévenin voltage source where the coupling factor

$$k_R = \frac{l_m}{l_R}$$ (5.12)

is additionally introduced. Fig. 5.4 shows the corresponding Thévenin and Norton equivalent circuits of the reduced order DFG model for connecting the model to the grid in simulation systems.

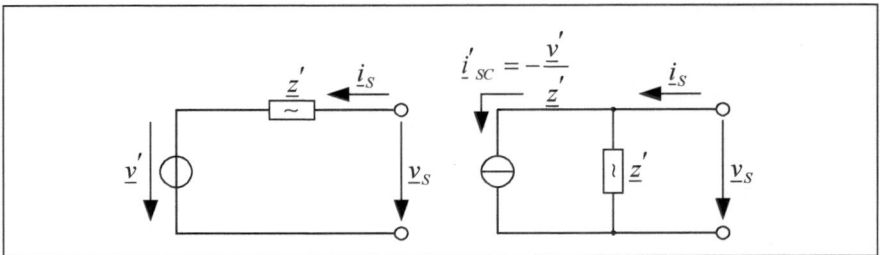

Fig. 5.4 Thévenin and Norton equivalent for coupling with the grid.

Note that the voltage source $\underline{v}'$ is, in general, variable and can be calculated by solving the rotor flux differential equations (5.13).

$$\frac{d\underline{\psi}_R}{dt} = -\frac{r_R}{l_R}\underline{\psi}_R - j(\omega_0 - \omega_R)\underline{\psi}_R + k_R r_R \underline{i}_S + \underline{v}_R \tag{5.13}$$

This equation can be derived from (5.2) by replacing the rotor current $\underline{i}_R$ with

$$\underline{i}_R = \frac{\underline{\psi}_R}{l_R} - k_R \underline{i}_S \tag{5.14}$$

which is derived from (5.4). For the stator current, the following expression can be deduced from (5.8)

$$\underline{i}_S = \frac{\underline{v}_S}{\underline{z}'} - \frac{\underline{v}'}{\underline{z}'} \tag{5.15}$$

The complete DFG model characterized by (5) - (12) can be described by a control schema shown in Fig. 5.5 where the following substitutions are introduced:

$$T_R = \frac{l_R}{r_R} \tag{5.16}$$

and

$$\Delta\omega = (\omega_0 - \omega_R) \tag{5.17}$$

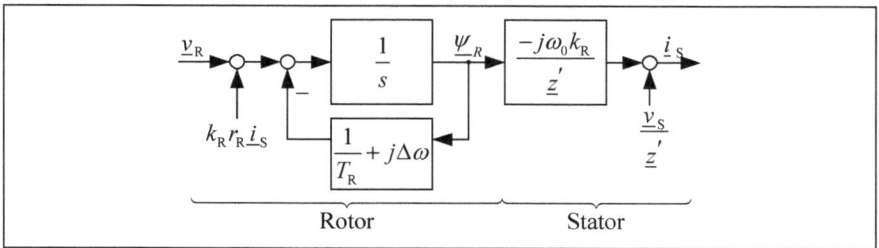

Fig. 5.5   Block diagram of the DFG reduced order model.

5.2.3   MSC Control Representations

A common approach for the control of DFG systems is the field oriented control as described in [17], [57] and [58]. Usually an outer power control loop works together with an underlying current control loop. In this approach an outer power control loop provides the current reference settings for the underlying active and reactive current controllers. A different approach using has been applied in [48] using a direct control of active and reactive power without an inner current control loop. The author suggests that this approach makes a more efficient use of existing hardware because the power control loop can be operated with smaller time constants than it would be possible in the case of separate power and current control loops that have different time constants to avoid interference.

The current controller is developed based on (5.13) by assuming steady state conditions, i.e. $d\underline{\psi}_R/dt = 0$. Then the corresponding rotor voltage can be directly calculated using

$$\underline{v}_{R,ref}^{D} = \left(\frac{1}{T_R} + j\left(\omega_0 - \omega_R \right) \right) \underline{\psi}_R - k_R r_R \underline{i}_S \tag{5.18}$$

To underline that the required rotor voltage $\underline{v}_{R,ref}$ may differ slightly from that calculated by (5.18) the superscript "D" is introduced indicating the direct steady state contribution. The control contribution of $\underline{v}_R^{D}$ is similar to that of a feed-forward controller. However, in real applications it is usually calculated by using the instant values of rotor flux and stator current instead of the corresponding reference values. The entire rotor voltage reference to be generated by the MSC is determined by extending the feed-forward term by the output of an additional current controller $\underline{v}_{R,ref}^{C}$.

$$\underline{v}_{R,ref} = \underline{v}_{R,ref}^{D} + \underline{v}_{R,ref}^{C} \tag{5.19}$$

using

$$\underline{v}_{R,ref}^{C} = k_{C,R} \left(1 + \frac{1}{sT_{C}}\right) \cdot \Delta\underline{i}_{R} \tag{5.20}$$

with

$k_{C,R}$: proportional gain of the rotor current control

T_{C} : integral time constant of the current control

and

$$\Delta\underline{i}_{R} = \underline{i}_{R,ref} - \underline{i}_{R} \tag{5.21}$$

The task of this controller with usually PI characteristic is on the one hand to compensate for parameter and measurement uncertainties and on the other hand to improve the dynamic control behavior.

As the direct feed-forward control contribution results in decoupling of the control channels (note that complex quantities are used for deriving the model) the additional current controllers according to (5.20) will control the real and imaginary rotor current components independently.

For multi-megawatt turbines, the stator resistance is very small, and thus negligible for the controller design. Then the relation between stator and rotor currents under steady state conditions can be written by using (5.1) and (5.3) as

$$\underline{i}_{R} = -\frac{l_{S}}{l_{m}}\underline{i}_{S} - j\frac{\underline{v}_{S}}{l_{m}} \tag{5.22}$$

Under the simplified conditions that the stator voltage is considered to be constant, the rotor current deviation can be calculated as

$$\Delta\underline{i}_{R} = -\frac{l_{S}}{l_{m}}\left(\underline{i}_{S,ref} - \underline{i}_{S}\right) \tag{5.23}$$

where the stator reference current $\underline{i}_{S,ref}$ is introduced. For the rotor current controller then can be written

$$\underline{v}_{R,ref}^{C} = k_{C,S} \left(1 + \frac{1}{sT_{CI}}\right) \cdot \left(\underline{i}_{S,ref} - \underline{i}_{S}\right) \tag{5.24}$$

with

$$k_{C,S} = -\frac{l_{S}}{l_{m}} k_{C,R} \tag{5.25}$$

The control schema of the MSC based on the equations discussed above is shown in Fig. 5.6. The converter itself is represented by a first order delay block.

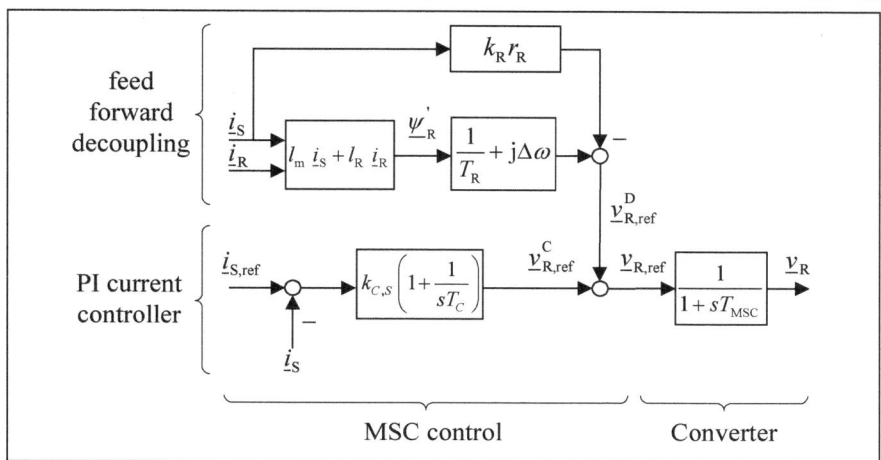

Fig. 5.6 Simplified current control representation using stator current control.

The rotor flux $\underline{\psi}_R$ required in Fig. 5.6 can be calculated from (5.4). However, due to parameter and measurement uncertainties, it may differ slightly from the real rotor flux. Therefore it is denoted as $\underline{\psi}'_R$. In practical implementations usually observer based methods are used to determine the rotor flux providing more accurate results [48]. In the simulation model, however, parameter and measurement tolerances don't need to be considered.

5.2.4 Combined DFG and MSC Model

For better understanding of the following steps the DFG model of Fig. 5.5 and the MSC models shown in Fig. 5.6 are drawn in a single diagram shown in Fig. 5.7.

Taking into consideration that the time constant of the converter T_{MSC} is small in comparison with the delay caused by the generator the block representing the converter can be eliminated. With the assumption $\underline{\psi}'_R \approx \underline{\psi}_R$ it then becomes obvious that the feedback through $(1 / T_R + j\Delta\omega)$ in the DFG model and the corresponding term in the controller compensate each other and can be deleted. The same applies to the terms $k_R r_R \underline{i}_S$. This becomes visible if v_R in (5.13) is replaced by (5.18). Therefore, these terms can be deleted in the overall model. Further simplification is possible by incorporating the term $-j\omega_0 k_R / \underline{z}'$ into the PI controller gain. With the assumption $r_s \ll j\omega_0 l'$ and thus $\underline{z}' \approx jx' = j\omega_0 l'$ the control gain becomes:

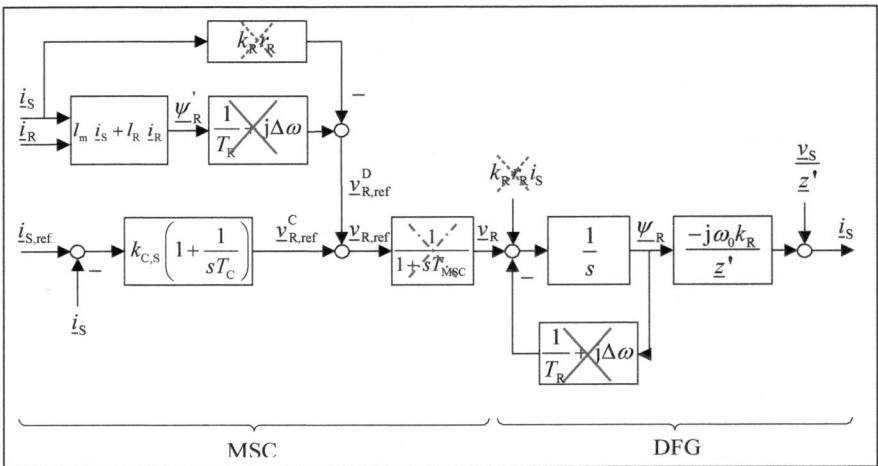

Fig. 5.7 Model of machine and rotor current control, and proposed changes in structure.

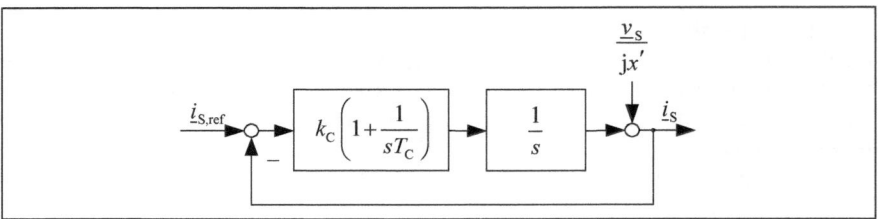

Fig. 5.8 Aggregated model of machine and rotor current control.

$$k_C = k_{C,S}\left(-\frac{j\omega_0 k_R}{j\omega_0 l'}\right) = -\frac{k_{C,S}k_R}{l'} = \frac{l_S}{l_R \cdot l'}k_{C,R} \qquad (5.26)$$

The simplification steps describe above lead to the aggregated MSC and DFG model shown in Fig. 5.8 for the MSC and generator.

5.2.5 LSC Model Representation

In a DFG, the active power of the turbine p_{WTG} that is fed into the grid is the sum of the stator power p_S and power of the line side converter p_{LSC}. For steady state operation and if losses are neglected, the active power of the LSC fed into the grid is defined by the stator power and the rotor speed [48],[58]

$$p_{LSC} = \frac{\omega_R - \omega_0}{\omega_0} p_S \qquad (5.27)$$

The total active power of turbine p_{WT} is the sum of the stator power p_S and power via the LSC p_{LSC}.

$$p_{WT} = p_S + \frac{\omega_R - \omega_0}{\omega_0} p_S = \frac{\omega_R}{\omega_0} p_S \qquad (5.28)$$

The active power control of the LSC usually bases on the control of the DC-link voltage [48]. The response time of the LSC can be further reduced by adding the MSC power, which corresponds with the rotor power, as reference to the LSC control [17]. The reactive power can be controlled independently within the voltage and current limits of converter. However, for economical reason it is more suitable to generate reactive power in steady state operation by the MSC through the DFG than with the LSC. Following grid faults, however, the LSC can help, within the available current margin to inject reactive current to the grid even faster.

The representation used for the LSC is shown in Fig. 5.9. Usually a filter is also part of the LSC. But the capacity of the filter can be included in the definition of the current limits, therefore the capacity of the filter does not have to be modeled separately.

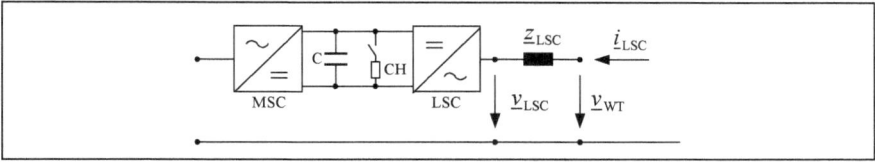

Fig. 5.9   LSC design.

As an approximation, the LSC can be seen as a fast acting voltage source (Thévenin equivalent) with a small delay described as first order lag. Using a Norton equivalent for describing the LSC (see Fig. 5.10), the current source is described by

$$\underline{i}'_{SC} = -\frac{\underline{v}_{LSC}}{\underline{z}_{LSC}} \qquad (5.29)$$

with

$\underline{v}_{LSC}$: voltage injected by the LSC

$\underline{z}_{LSC}$: LSC choke impedance

LSC Control Representation

For the further analysis of the LSC, an active current reference from the turbine control is assumed. This can be composed of the MSC power and the output of

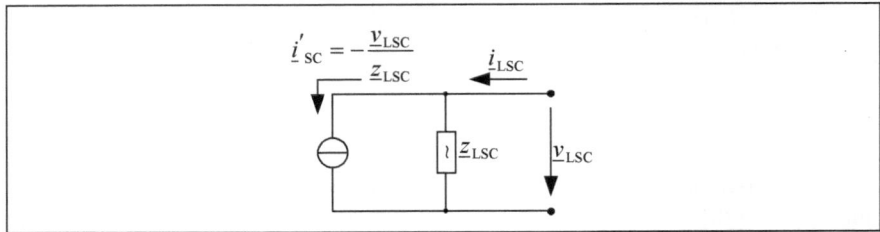

Fig. 5.10 Norton source representation.

the DC-link voltage controller. Besides, it is assumed that the DC-link voltage remains always within an operating range that allows a continuous operation of both MSC and LSC. For this purpose in modern wind turbines an additional DC-link energy absorber (DC-link chopper) is utilized. The active current reference can be calculated by dividing the LSC active power reference by the grid voltage. Further background of DC-link voltage control and DC-link protection to limit the DC-link voltage can be found in [56] and [48]. The common approach is to control the currents in a rotating reference frame is described in [17].

The LSC voltage $\underline{v}_{\text{LSC,ref}}$ is determined as the sum of two components:

$$\underline{v}_{\text{LSC,ref}} = \underline{v}_{\text{LSC,ref}}^{\text{D}} + \underline{v}_{\text{LSC,ref}}^{\text{C}} \tag{5.30}$$

where $\underline{v}_{\text{LSC,ref}}^{\text{D}}$ represents the feed-forward control part calculated for the steady state conditions

$$\underline{v}_{\text{LSC,ref}}^{\text{D}} = \underline{z}_{\text{LSC}} \underline{i}_{\text{LSC,ref}} + \underline{v}_{\text{G}} \tag{5.31}$$

and PI controller part $\underline{v}_{\text{LSC,ref}}^{\text{C}}$ to compensate for parameter and measurement uncertainties and on the other hand to improve the dynamic control behavior:

$$\underline{v}_{\text{LSC,ref}}^{\text{C}} = k_{\text{CP}} \left(1 + \frac{1}{sT_{\text{CI}}} \right) \cdot \left(\underline{i}_{\text{LSC,ref}} - \underline{i}_{\text{LSC}} \right) \tag{5.32}$$

with

$\underline{i}_{\text{LSC,ref}}$: grid reference current of LSC

k_{CP} : proportional gain of the current control

T_{CI} : integral time constant of the current control

The structure of the LSC model using a PI-control and decoupling for the LSC with its components controller, converter and Norton source is shown in Fig. 5.11.

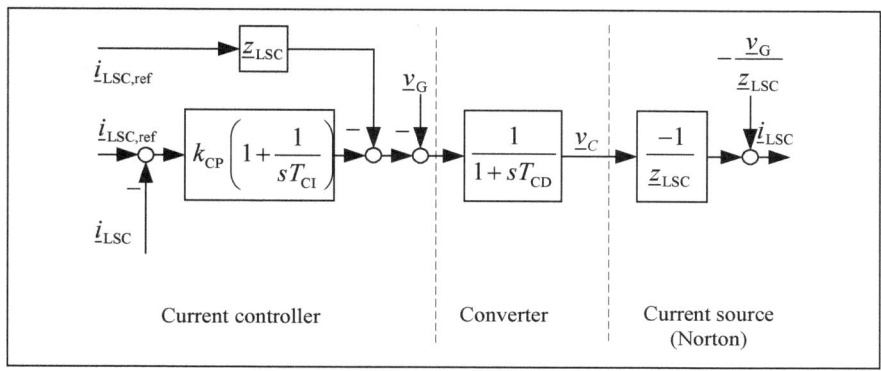

Fig. 5.11 Block diagram of current controller, converter and Norton source.

By moving the impedance $\underline{z}_{LSC}$ from the Norton source into the controller, we obtain the following slightly modified structure (Fig. 5.12).

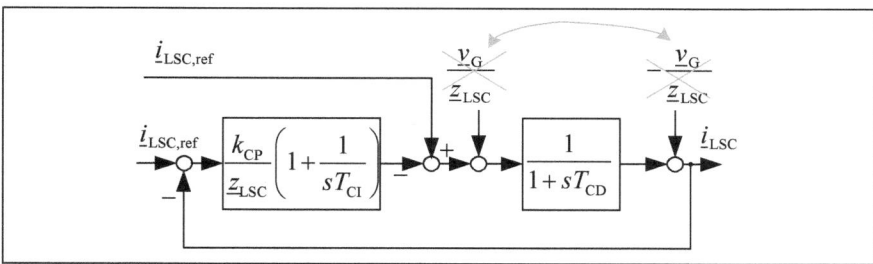

Fig. 5.12 Modified block diagram of current controller, converter and Norton source.

The time constant T_{CD} represents the delay of the converter which is very fast. Due to this fact the model can be further simplified as shown in Fig. 5.13.

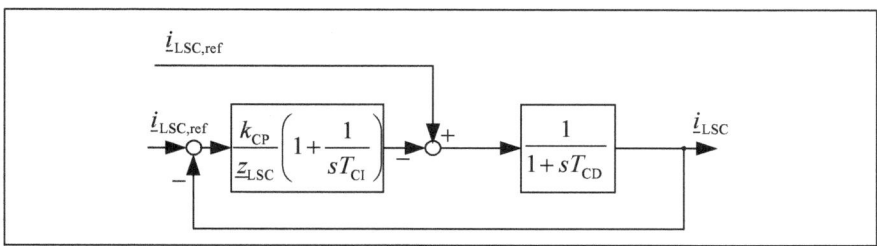

Fig. 5.13 Approximated model structure of the LSC.

The contribution of the PI-control is usually small compared to the feed-forward-control, therefore reducing the model of the LSC to a first order lag with a modified time constant T^*_{CD} (see Fig. 5.14) will usually yield a satisfactory response.

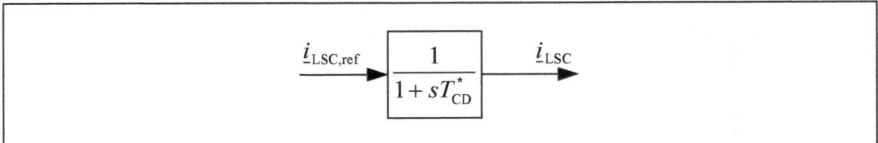

Fig. 5.14 Aggregated model structure of the LSC.

5.2.6 Aggregated DFG Model

Based on Fig. 5.8 and Fig. 5.14, and (5.28) an aggregated model for the DFG is proposed that includes MSC, DFG and LSC in a single structure in Fig. 5.15.

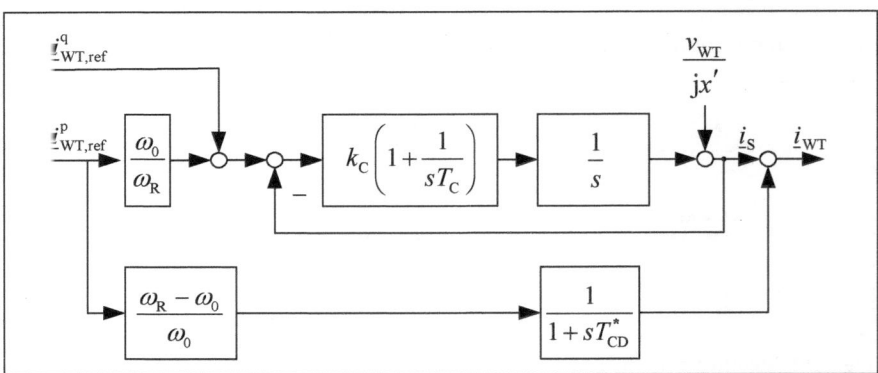

Fig. 5.15 Block diagram of current controller, converter and Norton source.

In general the LSC response is faster than that of the MSC supplying through the DFG to the grid. Therefore, in a simplified generic model the contribution of LSC can be considered by the steady state equations without modeling the LSC controller in detail. Due to the uncertainties concerning the reactive current injection in real WT implementations during grid faults and in order to develop a simple generic model the resulting WT current is derived from the stator current according to (5.33).

$$\underline{i}^{PQ}_{WT} = i^P_{WT} - j i^Q_{WT} \approx \frac{\omega_{R,n}}{\omega_0}\left(i^P_S - j i^Q_S\right) \tag{5.33}$$

For the active current Eq. (5.33) follows directly from (5.28). Additionally, it is assumed that $\omega_R = \omega_{R,n}$. As it can be seen in Fig. 2.4 in section 2.3, modern multi-megawatt turbines operate at or close to rated speed for operation points beyond 0.4 p.u. rated power. With this assumption the WT model becomes independent of the operating point.

Concerning the reactive current the assumption in (5.33) represents an acceptable simplification taking into consideration that the mismatch will be compensated by the PI current controller. Besides, the reactive current reference in steady state is usually small. The modified structure including DFG, MSC and LSC is shown in Fig. 5.16.

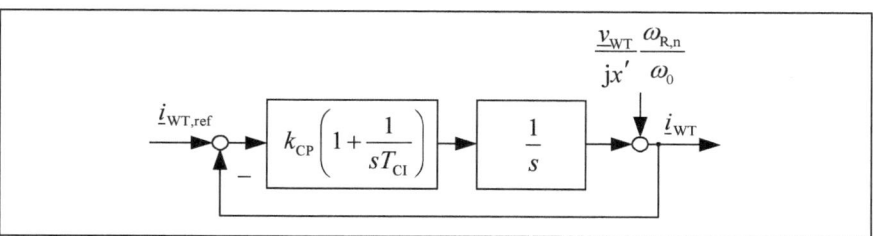

Fig. 5.16 Aggregated model of machine, MSC and LSC.

Selecting $\omega_{R,n}$ as rated speed may slightly underestimate dynamic currents at low power, but due to the reduced flux the error is also smaller than at rated power. Consider that the factor $\omega_{R,n}/\omega_0$ will affect only the current contribution associated with the transient reactance. Therefore a new equivalent reactance can be introduced as follows:

$$x_E' = x' \frac{\omega_{R,n}}{\omega_0} \tag{5.34}$$

The resulting structure for the aggregated model of the DFG shown in Fig. 5.17 is then achieved.

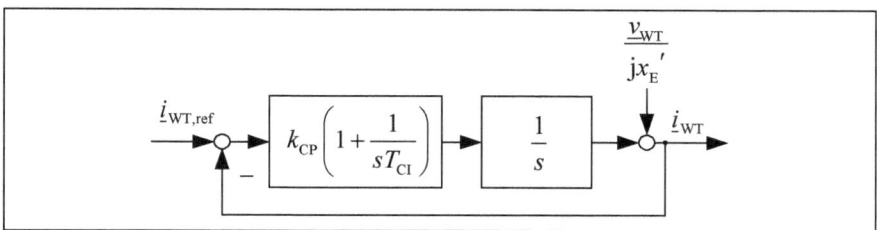

Fig. 5.17 Aggregated model of machine and current control.

5.2.7 Selection of Reference Frame

For the derivation of the DFG equations (5.1) - (5.4), only the rotational speed of the coordinate system has been defined by assuming grid synchronous frequency ω_0 but not the right position of the dq axes. In a wind turbine control system it makes sense to separate between active and reactive current controllers. This is achieved by defining the d axis along the terminal voltage. In such a coordinate system the active current will always correspond with the real part of the stator current and the reactive current with the reverse (negative sign) of the imaginary component. Terminal voltage orientation is therefore applied in this work for the definition of the control reference frame.

The DFG model is operated in a grid synchronous coordinate system, this is necessary to correctly represent the impact of voltage amplitude and voltage phase angle changes. The DFG and MSC model with the required coordinate transformations to establish active and reactive current control is shown in Fig. 5.18.

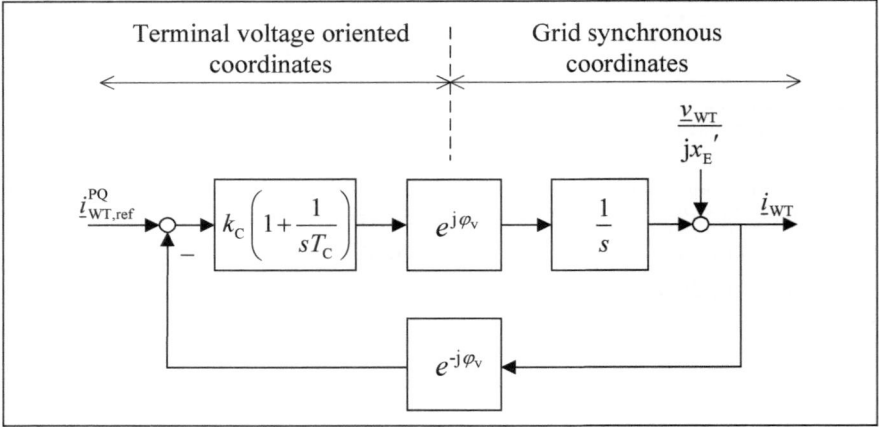

Fig. 5.18 Aggregated model of DFG and corresponding MSC controller including reference frames.

The angle φ_V represents the position of the terminal voltage in grid synchronous coordinates and is determined usually by a so called PLL (phase-locked loop) unit. In a simulation environment this angle is obtained by solving the grid equations.

The current $\underline{i}_{WT,ref}^{PQ}$ is defined as

$$\underline{i}_{WT,ref}^{PQ} = \underline{i}_{WT,ref}^{P} - j\underline{i}_{WT,ref}^{Q} \tag{5.35}$$

with the reference values for stator active and reactive currents

$$\underline{i}_{WT,ref}^{P} = \frac{p_{WT,ref}}{|\underline{v}_{WT}|} \tag{5.36}$$

$$i^{Q}_{\text{WT,ref}} = \frac{q_{\text{WT,ref}}}{\left|\underline{v}_{\text{WT}}\right|} \tag{5.37}$$

5.2.8 Optional Identification of Model Parameters

The models described in the previous sections are based on simplifications and aggregations. That means the corresponding generator parameter do not necessarily represent real WT parameters but rather approximations. Converter control parameters and limits will usually be deduced from parameters of the real system.

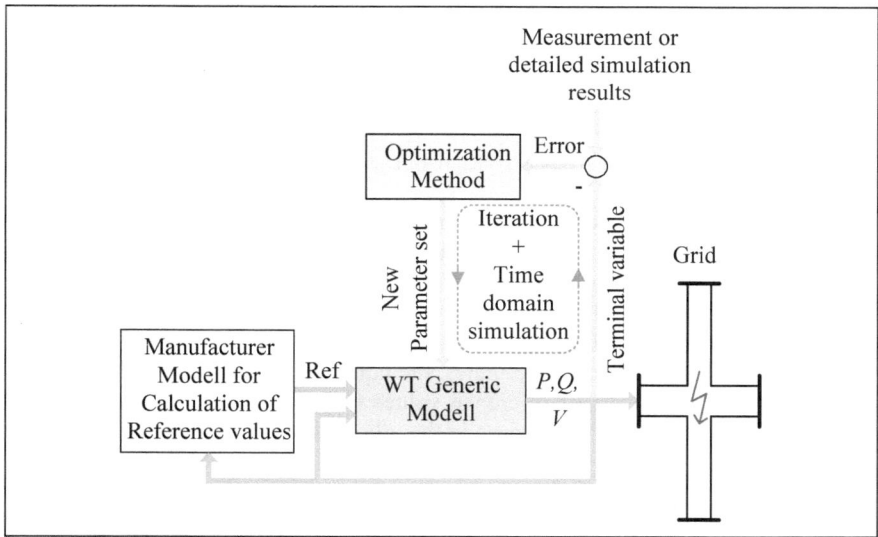

Fig. 5.19 Parameter identification procedure.

Generator and converter parameters like (k_{CP}, T_{CI}, x_{E}) can be deduced from physical parameters, but an improved model accuracy may be achieved by optimizing these values using identification techniques. The identification procedure is illustrated in Fig. 5.19.

As reference for comparison with the generic model measurements or accurate simulation results from models validated by measurements can be used. Those results should cover typical contingencies for which the model is intended to be used. It is also recommended to carry out the identification for several scenarios simultaneously. This will guarantee that the model parameters are not only tuned for a single case providing good accuracy but also for other scenarios that may be poorly represented.

5.2.9 Simplified Aggregated Model of DFG using First Order Lag

A further simplification of the DFG model is possible if the impact of the transient reactance is neglected [4]. The resulting structure (shown in Fig. 5.20) is then equivalent to the structure of the LSC (Fig. 5.14). Omitting the impact of the transient reactance does not change the steady state behavior, but it will have an impact on the response directly following transient events like voltage magnitude or voltage phase angle changes.

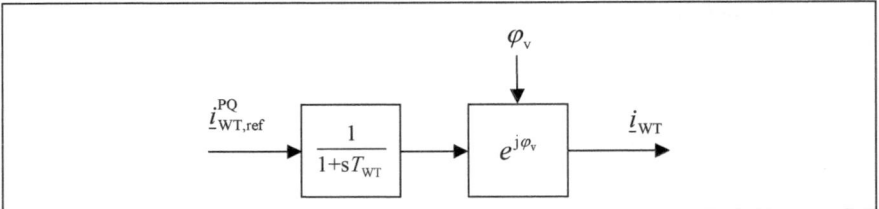

Fig. 5.20 Simplified aggregated model structure of DFG machine and current control.

The model is based on the assumption that the WT response is faster than the response of other power system components and the transient period not correctly represented in this model will not significantly affect the phenomena under investigation. But as will be shown later (see section 5.5) neglecting the transient impedance will lead to voltage spikes during fault clearance.

The model has only one parameter, the equivalent time constant T_{WT}. However, to calculate the correct reference values the models described in the next section have to be used, as is the case in the other more detailed models too.

5.3 Model of the FSC

The FSC has been used in wind turbines for more than two decades. A wide variety of generators exist at the moment, using synchronous wound rotors, permanent magnet synchronous machines or asynchronous machines, with gearbox, with a higher number of poles and a reduced gearbox and without gearbox. A typical structure is shown in Fig. 5.21. But from a grid connection point of view, the selection of the generator and the machine side converter does not influence the behavior towards the grid if MSC and LSC are decoupled by a DC link as it is usually the case in modern wind turbine designs. The generator and MSC are therefore represented as controllable current source only.

FSC Model Representation

As an approximation, the FSC can be seen as a fast acting voltage source (Thévenin equivalent) with a small delay described as first order lag. The impedance

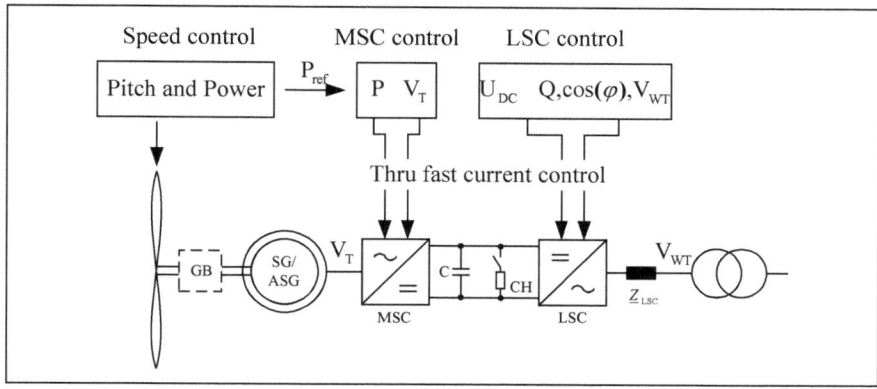

Fig. 5.21 Key components of FSC system.

$\underline{z}$ (corresponding with the transient impedance $\underline{z}'$ of DFG based system) can be defined as

$$\underline{z} = \underline{z}_{LSC} + \underline{z}_{Tr} \tag{5.38}$$

with

$\underline{z}_{LSC}$ = LSC choke impedance

$\underline{z}_{Tr}$ = transformer impedance

Using a Norton equivalent for describing the FSC (see Fig. 5.22), the current source is described by

$$\underline{i}'_{SC} = -\frac{\underline{v}_{WT}}{\underline{z}} \tag{5.39}$$

with

$\underline{v}_{LSC}$: Voltage injected by the LSC

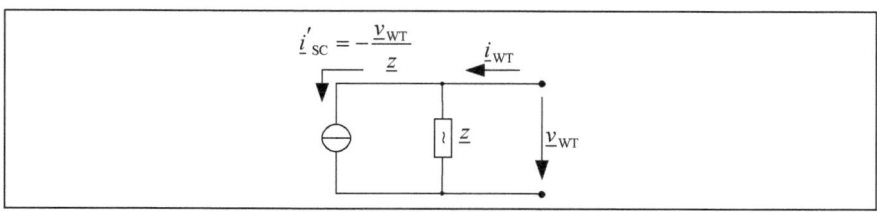

Fig. 5.22 Norton source representation.

Structure and control design of the FSC are identical to the design of the LSC of a DFG. Therefore the same structure as for the LSC in Fig. 5.14 and the resulting diagram in Fig. 5.20 can be used.

5.4 Model Structure

The model aggregation for the DFG in Fig. 5.17 represents models for generator, converter hardware and control. The model aggregation for the FSC in Fig. 5.14 represents the equivalent model for converter hardware and control, the generator itself is represented as a controllable current source only.

The block diagram in Fig. 5.23 shows an implementation of the generator model using as input active and reactive current reference and turbine voltage and as output real and imaginary parts of the current (in grid coordinates).

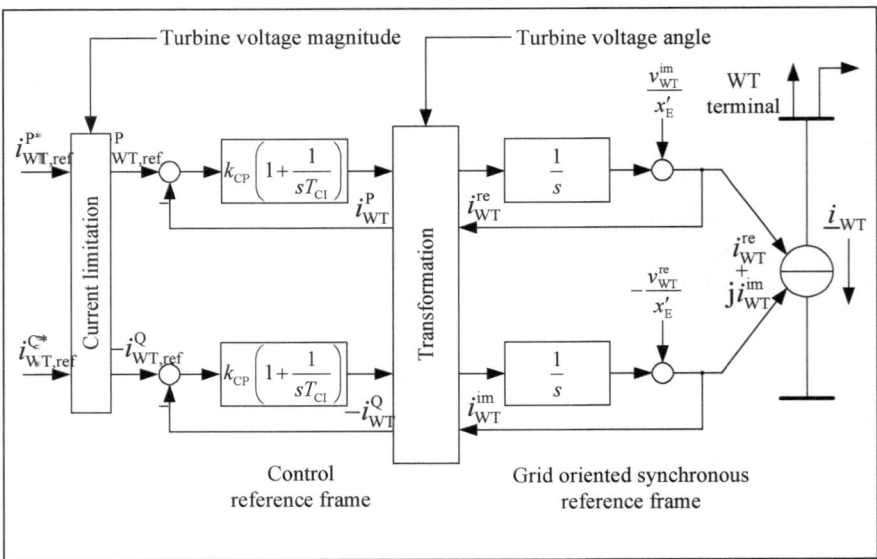

Fig. 5.23 Core of aggregated generator model of machine and current control as block diagram including terminals.

5.4.1 Implementation of the Norton Equivalent (Current Source)

In stability kind of power system simulation software, the grid is usually described by complex algebraic equations. Therefore, following grid faults the voltage magnitude and phase angle can change immediately. To improve the interaction of the WT model with the grid and thus improve the convergence behavior of the simulation, it is recommended to incorporate the term "stator voltage divided by the transient reactance" into the grid equations. The necessary modification is shown in Fig. 5.24.

The stator current is fed back to the transformation as real and imaginary current. Fig. 5.25 shows an implementation that can be used if the current of the Norton source is not directly available for the simulation model. Compared to Fig. 5.23 the summation point for calculating real and imaginary currents (i_{WT}^{re} and i_{WT}^{im}) that adds the impact of the impedance x_E' is moved to the feedback loop. All three implementations (Fig. 5.23 - Fig. 5.25) are numerically identical in steady state conditions.

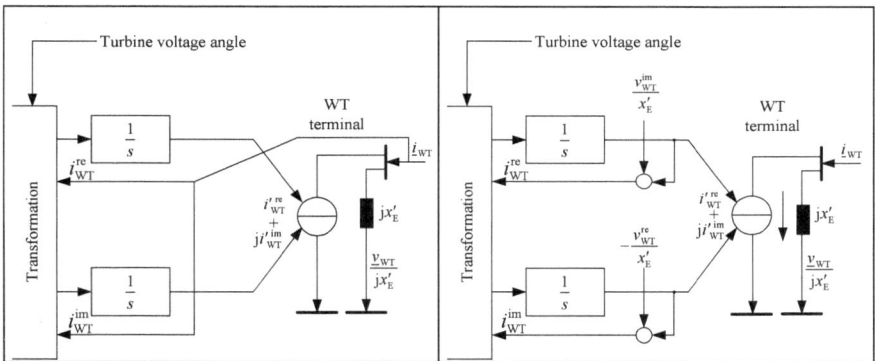

Fig. 5.24 Coupling of WT model with the grid via current source model.

Fig. 5.25 Coupling of WT model with the grid via current source model in case the currents are not available directly.

5.4.2 Current Limitation

The output current limitation of the generator model is shown in the current limitation block in Fig. 5.26. The current limitation can be divided into three main categories

(1) control limits

(2) dynamic limits

(3) physical limits

Control Limits

Active and reactive current output limitations can often be represented as function of the voltage. Although the actual implementation may be different, a good

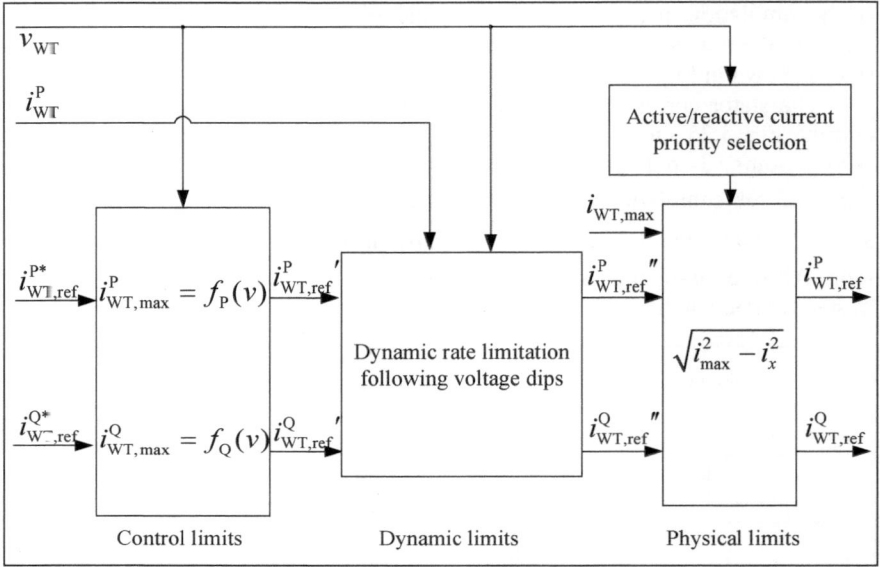

Fig. 5.26 Current limitation block overview.

approximation can usually be achieved by defining a table that limits active and reactive current as function of voltage:

$$i_{WT}^{P}{}' = \max\left(i_{WT}^{P*}, f_{P}(v_{WT})\right) \quad \text{and} \quad \left|i_{WT}^{Q}{}'\right| = \max\left(\left|i_{WT}^{Q*}\right|, f_{Q}(v_{WT})\right) \tag{5.40}$$

with i_{WT}^{P*} and i_{WT}^{Q*} as active and reactive current reference values from the turbine controller. The main reasons for implementing an active current limitation as function of voltage are

(1) implementing a reactive current priority during grid faults,

(2) active current limitation to fulfill specific grid codes

(3) grid stability limits at weak grid connection points [57].

Reasons for reactive current limitations as function of voltage are mainly

(1) implementing specific converter limits

(2) emulating a specific reactive current control during grid faults.

Dynamic Limits

Dynamic control limits are needed to allow an improved representation of active power recovery after grid faults. Following grid faults, a slow recovery of active and reactive power or currents may be required. These currents are described by $i_{WT,ref}^{P}{}''$ and $i_{WT,ref}^{Q}{}''$ in Fig. 5.26. Common reasons are

(1) limiting shaft and gearbox loads

(2) limiting DC-link voltage variations due to very fast active power changes.

(3) limiting voltage changes to prevent overvoltage following grid faults

Depending on the manufacturer, this function may be implemented as first order lag or as linear rate ramp limitation. Therefore a model should preferably allow both methods.

In addition, there may be a delay after the voltage starts rising again until the active power increases. This can be due to control delays in the converter, a settling time needed by the PLL to synchronize again after voltage dips to very low voltages and as result of phase angle jumps after fault clearance. Generally, this delay is more relevant for very deep voltage dips.

Note that dynamic active current limits following voltage dips may be the result of an active power recovery function implemented a turbine control level (see section 4.5.1) or - possibly in addition – in the converter block.

Physical Limits

Short term thermal current limitations define physical limits for the maximum current the system can feed into the grid. These current limits of the DFG and FSC are always defined by the converter that has far shorter time constants than the generator. Commonly, active current is prioritized during steady state operation while a reactive current priority may be used during very low or high voltage conditions to reduce the voltage changes. Assuming the maximum WT current allowed is $i_{WT,max}$ the limitations are depending on the priority settings as follows for active and reactive current priority:

$$
\begin{aligned}
i_{WT,ref}^{P} &= i_{WT,ref}^{P}{}'' \\
i_{WT,ref}^{Q} &= \sqrt{\left(i_{WT,max}\right)^2 - \left(i_{WT,ref}^{P}{}''\right)^2}
\end{aligned}
\quad \text{or} \quad
\begin{aligned}
i_{WT,ref}^{Q} &= i_{WT,ref}^{Q}{}'' \\
i_{WT,ref}^{P} &= \sqrt{\left(i_{WT,max}\right)^2 - \left(i_{WT,ref}^{Q}{}''\right)^2}
\end{aligned}
\quad (5.41)
$$

Converters are often designed to have a very short (several 100 ms) overloading capability that allows to provide higher currents for a limited time period, especially during faults. This can be modeled by defining different values for for the maximum current in normal operation ($i_{WT,max}$) and a higher value during grid faults $i_{WT,max,FRT}$).

5.4.3   Wind turbine Transformer

The objective of a generic model representation is to achieve a reasonable approximation of reality using a simplified model structure. Currently different topologies exist on the market, controlling active and reactive power either on the grid side or on the turbine side of the transformer.

One possibility would be to add a separate transformer model in the case of control at the turbine side (LV) terminals. But this would make it more difficult to use this generator model as part of an aggregated wind plant. The impact of the transformer can be described as in (5.42)

$$\underline{v}_{\text{LV}} = \underline{v}_{\text{MV}} - \left(r_{\text{Tr}} + jx_{\text{Tr}} \right) \underline{i}_{\text{MV}} \tag{5.42}$$

with r_{Tr} and x_{Tr} as resistance and impedance of the transformer.

By including the transformer equations (5.42), a simple transformer model allows to include the transformer in the case of low voltage control or - by setting the two parameters to zero, to take out the transformer in the simulation model in the case of control at the MV terminals.

5.4.4   DC-Link Energy Absorber for FSC

In the case of grid faults, the energy that can be transferred into the grid is limited both for physical reasons and due to the limited current capability of converters. Since the energy provided by the wind does not necessarily change, a grid fault with voltages below 90% rated voltage at rated power will lead to an increase of rotor speed. By changing the blade pitch angle, the power input from the wind can be reduced, but the time constants of the pitch system are long compared to the duration of grid faults. Unless additional measures are taken, the rotor speed of a turbine with FSC will increase in a comparable way to a turbine with DFG.

FSC wind turbines are commonly equipped with a controlled resistor in the DC-link ("DC-link chopper") or a comparable technology to absorb at least a part of the energy provided by the generator that cannot be transferred to the grid in the case of grid faults. A small rating of this device will only absorb limited current following a voltage drop, as a result the turbine rotor speed will increase during a fault.

A large rating of the DC-link chopper can allow to absorb up to rated turbine power, in this case the rotor speed would not increase even if the voltage drops down to zero. The mechanical power is always equal to the power reference ($p_{\text{mech}} = p_{\text{ref}}$), even though the electrical power p_{el} is reduced during grid faults.

5.4.5 Resulting Generator and Converter Model

The resulting generator model including current limitation and transformer for the DFG model is shown in Fig. 5.27, the equivalent model including DC-link energy absorber for the FSC is shown in Fig. 5.28. This approach does not exactly represent the configuration in real turbines where functionality may be implemented either in the wind plant control, the turbine control or the converter. But it allows a modular model design. Typical functions that are part of the electrical control block are the calculation of the active current reference as function of external power reference, rotor speed (see section 4.5.1). The calculation of the reactive power reference is described in chapter 6.

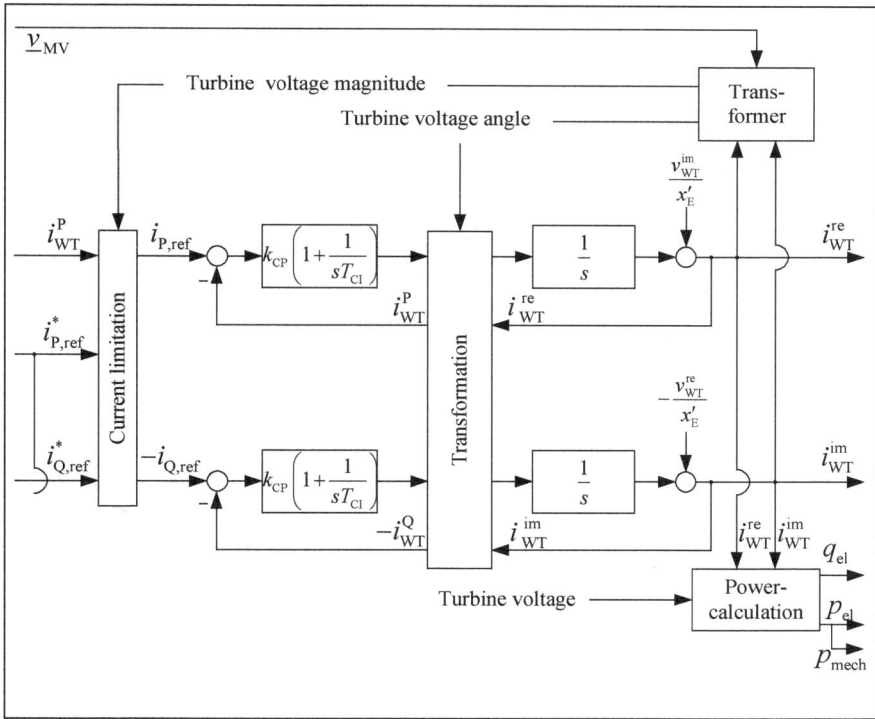

Fig. 5.27 Detailed block diagram of aggregated generator model for DFG and FSC.

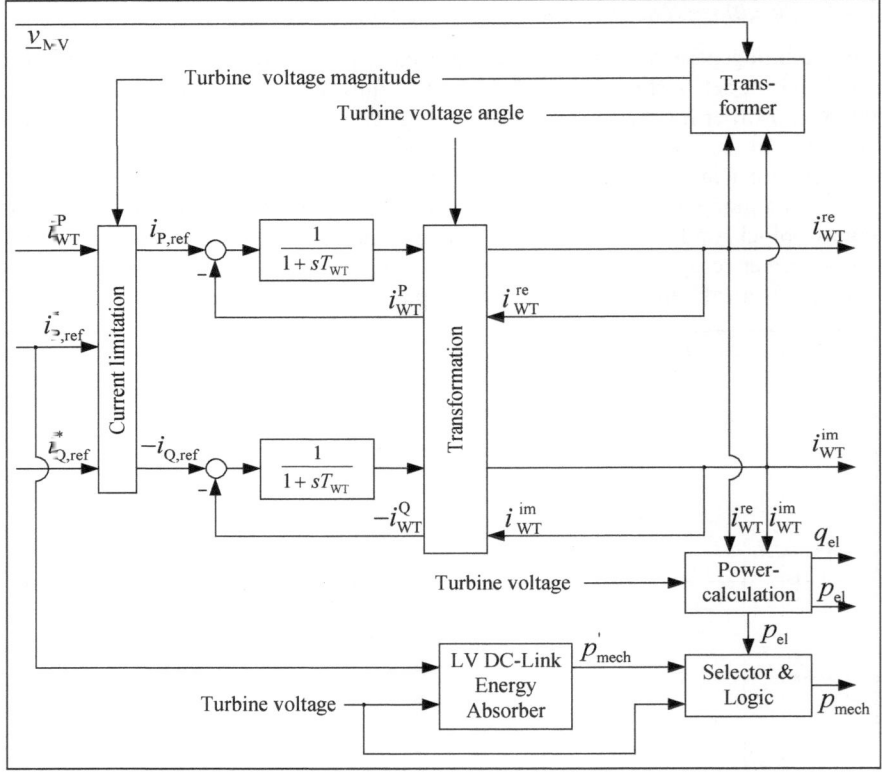

Fig. 5.28 Detailed block diagram of aggregated generator model for FSC.

5.5 Validation Results

A measurement setup according to [59] and [60], (see also [53] and Fig. A.3) has been used to compare the results of simulations with measurements for a 2 MW DFG and a 4.5 MW FSC based wind turbine. The producer oriented sign convention is applied for all validation figures, generator active power and overexcited reactive power are considered to be positive.

5.5.1 Measurements and Simulations of DFG based Wind Turbines

The turbine simulation model is based on the descriptions of sections 2 - 4 with the proposed generator model in section 5. For the comparison, three-phase measurement data is decomposed into positive and negative sequence components according to [59]. The output of the simulation is filtered with 50 Hz in order to have an equivalent to the positive and negative sequence values of measurements [53].

Measurements of a voltage dip down to 20% of the rated voltage of a DFG are compared with simulations in Fig. 5.29 for a 2 MW wind turbine. The reference values for the converter of active power have been taken from measurements. This is necessary to evaluate the accuracy of the generator and converter model only and helps to avoid showing uncertainties of arising from both the aerodynamic, mechanical and control models of the turbine and uncertainties of the wind speed measurement on the reference values. The reactive current reach-

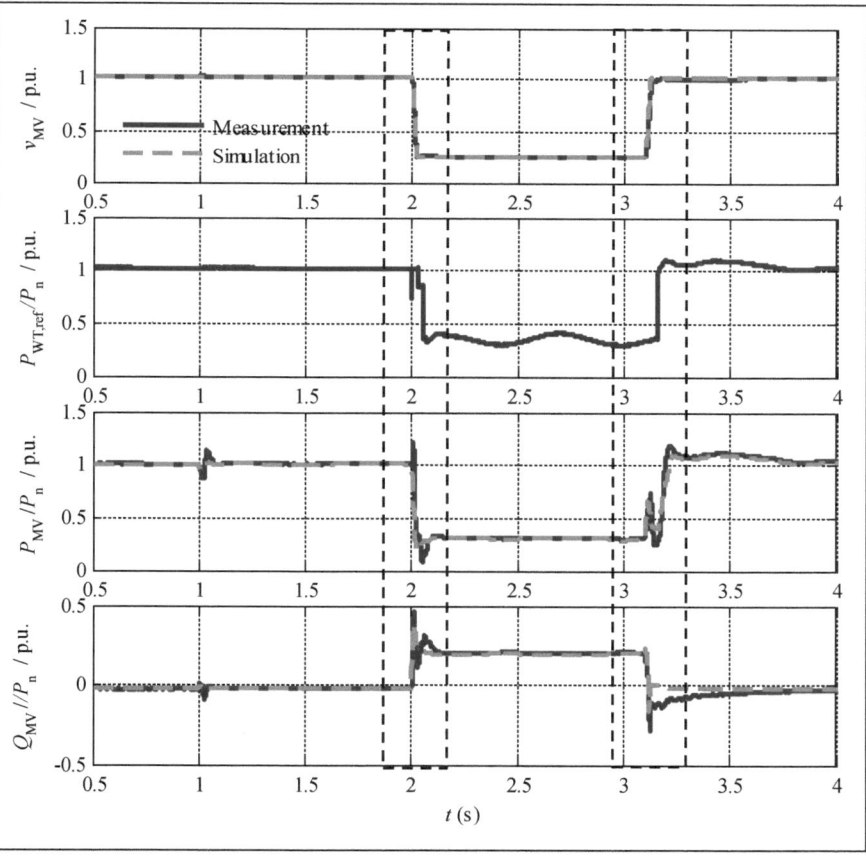

Fig. 5.29 Measurement and simulation of voltage, active power reference, active power and reactive power during a voltage dip down to 20% rated voltage for wind turbine with DFG operating at rated power. Results using the proposed aggregated model shown in Fig. 5.23 including a transformer are compared to MV measurements. A close-up-view of the dashed areas is shown in the following figures.

es ̇ p.u. during the fault in both measurement and simulation as required by German grid codes.

The simulation shows a very good correlation of both active and reactive power before, during and after the voltage dip. A difference in the reactive power following voltage recovery remains between the models and measurement due to the transformer inrush following voltage recovery at t = 3.2 s that cannot be modeled accurately using RMS simulation.

5.5 2   Comparison of Proposed and Simplified DFG Model

The moments of voltage drop and voltage recovery for a DFG from Fig. 5.29 are shown in Fig. 5.30. The measurements are compared with both the proposed aggregated model (as shown for example in Fig. 5.18) and to the simplified model according to Fig. 5.20.

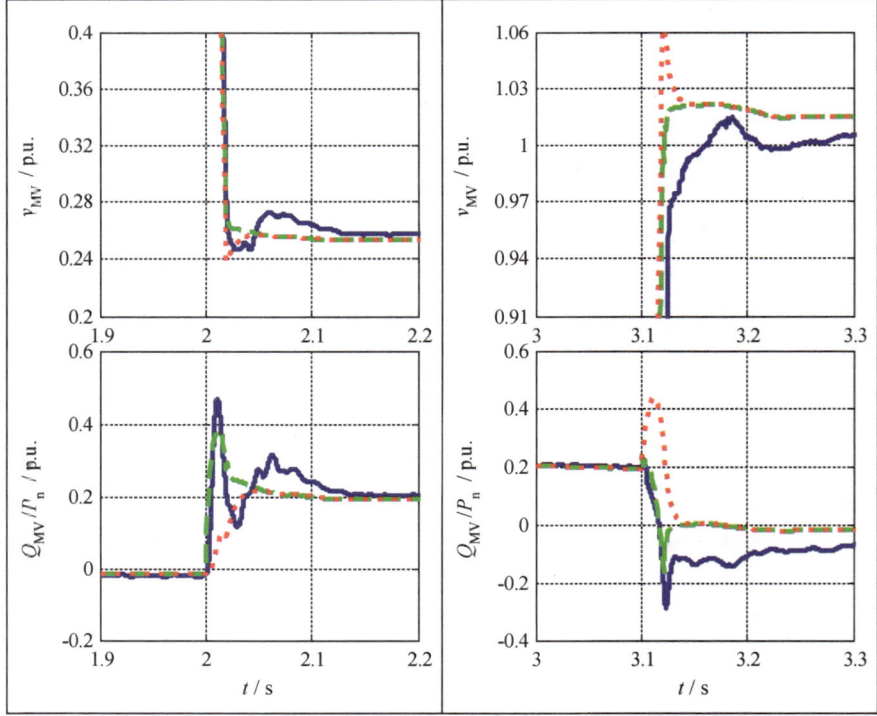

F g. 5.30 Comparison of measurement of DFG (solid) and simulation using the proposed generator model (dashed) and using the simplified first order lag model (doted) for voltage, active power and reactive power during voltage dip down to 20% rated voltage (left) and voltage recovery (right).

The figures show voltage and reactive power of a DFG based turbine at the beginning and at the end of a voltage dip. As can be seen, the aggregated model shows a far better representation, especially during the first 50 ms following the voltage drop and following the voltage recovery. During this period, the reactive power is modeled far more accurately.

Especially the reactive power spike following the voltage recovery of the simplified model (with an error of 0.5 p.u. compared to both measurement and aggregated model) can lead to voltage spikes that can cause stability problems during simulations by possibly triggering protection equipment action. By contrast, the proposed model (dashed trace) does not lead to any voltage overshoot.

5.5.3 Measurements and Simulations of FSC based Wind Turbines

Measurements of a voltage dip down to 20% of the rated voltage of a FSC are compared with simulations in Fig. 5.31 for a 4.5 MW wind turbine. The reference values for the converter active power and is set to a constant value as it would be expected for a type 4 control. The limitations of the active and reactive current as described in Fig. 5.26 lead to almost the same results as the measurements. The active power recovery is reproduced rather accurately by the proposed dynamic limiting block. The active power is held almost constant following the fault due the activation of the DC-link chopper during the fault which kept the rotor speed constant. This is reproduced well by the model.

The overshoot of voltage the during voltage recovery at t = 2.6 s by about 10% compared to the pre-fault value due to a slight delay between voltage recovery until the reactive current is reduced is modeled well. Fig. 5.32 shows that the reactive current is reproduced accurately.

5.5.4 Validation Results

Specific effects of the different technologies are reproduced very well. The DFG has an overshoot of the reactive current at the begin of the of the voltage drop. During voltage recovery, the reactive power becomes inductive and an overshoot of the voltage is avoided (see Fig. 5.30 at t = 3.12 s).

The FSC does not have an overshoot of the reactive current throughout the entire measurement (see Fig. 5.32). But due to a slight delay between voltage recovery and reactive current reduction, an overshoot of the voltage during voltage recovery is induced, (see also [61] on this an effect).

A comparison of Fig. 5.29 and Fig. 5.31 shows different active power recovery strategies, but both can be reproduced with the proposed model.

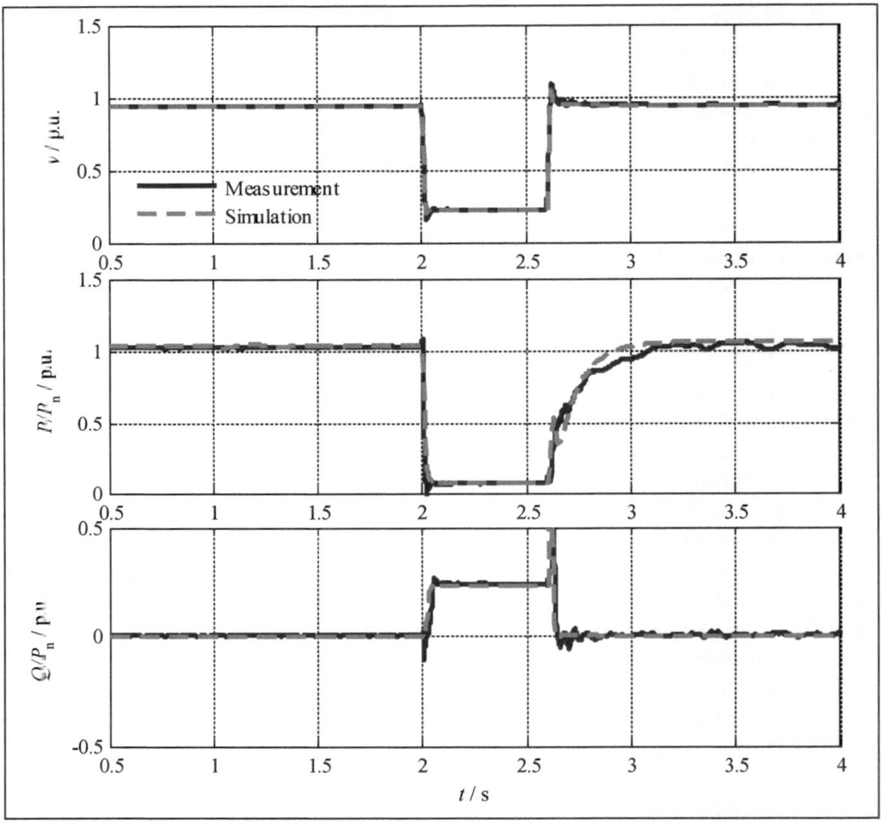

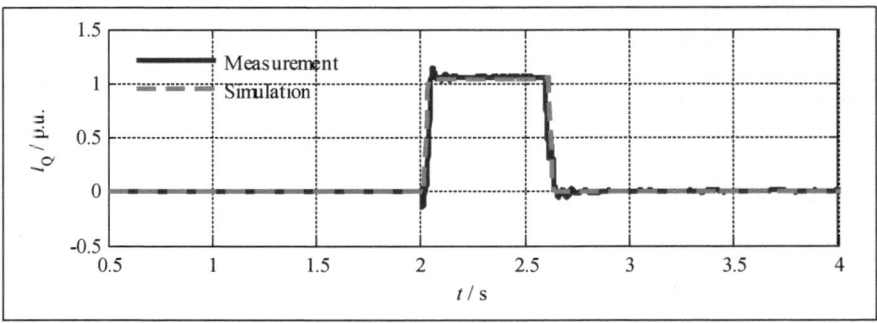

Fig. 5.31 Measurement and simulation of voltage, active and reactive power during a voltage dip down to 20% rated voltage for wind turbine with FSC using the proposed aggregated model shown in Fig. 5.27.

Fig. 5.32 Measurement and simulation of reactive current during event described in Fig. 5.31.

5.6 Summary

A generic generator model for both DFG and FSC systems is proposed. The model is derived from the physical equations of the generator, the converter hardware and converter control system. Therefore, the key parameters can be derived from available data sheet and controller information.

The model offers a significant reduction of complexity compared to detailed models with separate generator and converter models. It shows very good correlation with measurements of both with DFG and FSC MW-size wind turbines while still requiring only a very limited number of parameters. The model is not applicable for older DFG systems that require a crowbar during grid faults.

A key requirement for generator and converter models of turbines is the adequate representation of current limitation functions during grid faults and during steady state operation. Generic, manufacturer independent current limitation function have been presented and validated with measurements for DFG and FSC turbines.

The proposed model using a transient reactance and is compared to a simplified generator model based on a first order lag only. The proposed model improves the dynamic representation following fast voltage changes and avoids voltage spikes during voltage recovery compared to the existing generic models.

6 Reactive Power Control of Wind Plants

6.1 Introduction

The intention of this chapter is to describe a reactive power control structure for wind turbines and wind plants that provides the necessary capability to ensure a stable system voltage in the medium and high voltage grid even if no synchronous generation is present any more.

Grid codes started to consider renewable sources of energy as a relevant contribution in Germany in 2001 [23] when requirements for FRT capability and reactive power control at plant level for MV and HV connected wind turbines were introduced. This addresses two key questions for the steady state operation of grids: provision of active power to keep the system frequency stable and provision of reactive power to support a sufficient voltage level.

Starting in 2003, the next relevant issue was addressed, the requirement to provide reactive current during faults to limit the impact of grid faults and following grid faults to ensure a recovery of the voltage to the pre-fault level [62]. The inherent voltage control capability of synchronous generators to provide reactive power during faults [63] was taken as reference for this requirement [64].

6.1.1 Limitations of the existing grid code requirements

Existing grid codes are usually based on the assumption that a significant number of synchronous generators remains operating. Even newer studies [65] assume a certain minimum number "must run units" , conventional power stations with synchronous generators that operate – if necessary even without active power production as synchronous condenser – to stabilize the voltage.

Key limitations of existing grid codes are:

■ The impact on the voltage of fully replacing synchronous generators by wind plants has not been analyzed systematically.

■ The requirements for a dynamic reactive current contribution for example in [66]-[68] had originally been specified for the high voltage connection point but are now applied to medium voltage or even in most cases to the low voltage terminals of wind turbines. As consequence the effective impact at high voltage level is considerably lower.

■ Deadbands in the dynamic reactive power control of wind turbines have been accepted to ease the use of older control concepts. But as consequence, the resulting effective current contribution at the high voltage terminals is reduced even further.

6.1.2   Typical wind plant configuration

Set points and control response of wind plants are defined with respect to the point of common coupling (PCC). The task of the wind plant controller (see Fig. 6.1 for a typical structure of a wind plant) is to determine the set points of the individual wind turbines (WT) and possibly of the additional active or passive reactive power compensation equipment, so that the required control objectives at the PCC are met.

A central control unit coordinates the reactive power control. It receives commands from the transmission or distribution system operator, measures voltages and currents at the medium or high voltage terminals and controls the turbines by setting the corresponding V/Q reference values. The requirements on Var control in steady state are generally not time-critical. The changes in the grid are due to load changes and/or line or load switching. For modifying the V/Q reference of the WT a time frame of some seconds to several minutes is usually sufficient.

A certain degree of coordination is necessary between different equipment that can influence voltage or reactive power. On-load tap changers (OLTC) of transformers may trigger a stepping within 7 s .. 30 s if the voltage during this time remains permanently outside a given limit. Since it is preferable to change the reactive power of the wind turbines first to control the voltage, it is evident

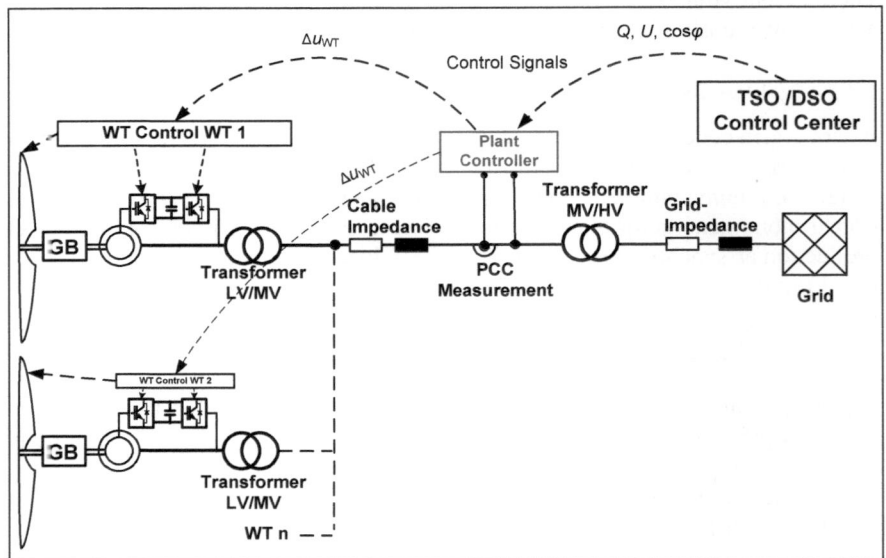

Fig. 6.1   Example of wind plant control structure.

that the WT must respond faster than the OLTC. Equivalent considerations apply for shunt reactors and capacitor banks.

The chapter consists of 4 parts.

■ In part 1, the general requirements for reactive power control of power stations and wind plants are summarized

■ In part 2, the capability of synchronous generators to dynamically stabilize the grid voltage is analyzed and requirements for wind plants are deduced

■ In part 3, based on the analysis, two alternative control structures for wind plants are deduced that provide a fast dynamic voltage stabilization.

■ In part 4, the developed control structure is compared to both synchronous generators and other existing control concepts for wind plants

Compared to the previous chapters which are confined to wind turbines and their components, the attention of this chapters is dedicated to the Wind Power Plant (WPP) as a whole and its major service provision regarding stability improvements.

6.2 Reactive power Requirements of Power Stations and Wind Plants

Reactive power control of wind turbines and wind plants started from power factor control of turbines as individual units, which dates back to the mid-90s. In 2001 E.ON Netz [23] issued a grid code that required reactive power control by a wind plant controller and an input (interface) for external set-points at the PCC.

As a reaction to the increase in wind power installations, E.ON Netz from 2003 onwards [62] also required a fast reactive current control in case of grid faults in addition to the slow reactive power control.

6.2.1 Reactive Power Control during Normal System Conditions

Reactive power control during normal system conditions usually requires a slow response only; common timeframes are 7 s .. 60 s ([66], [67]), which can go up to several minutes until a final value is reached. The following operation modes are common:

(1) power factor control, using φ $\cos(\varphi)$, $\tan(\varphi)$, including unity power factor

(2) reactive power control

(3) voltage control including voltage static / Q(V)-characteristic

The set point can be defined as a fixed value or preferably as a function of the operating point.

Power factor control is the most common requirement for wind plants. This is partly due to historical experiences. Power factor control is common in distribution systems for loads, and it requires no communication infrastructure for direct (online) control to the reactive power setpoint. This same reasoning explains why reactive power control is usually not used since it would need a frequent adaptation of the set point.

In order to respond to voltage deviations a voltage versus the required reactive power (voltage static, Q(V)) or alternatively power factor characteristic can be used. In this case, the plant controller can react to system voltage changes without the need for a centralized adaptation of a controller reference value.

6.2.2   Fast Voltage Control to Sudden Voltage Changes

Synchronous generators inherently stabilize the voltage by feeding in reactive currents within milliseconds as response to sudden changes in voltage. Since wind turbines begin to replace synchronous generators, providing some basic support for stabilizing the voltage when sudden changes of voltage occur is now required from wind turbines in most grids with high penetration of wind power ([24], [51], [52], [68] and [69]).

Fig. 6.2 shows an example of voltage control requirements by E.ON Netz [62]. This figure describes the required proportional gain of the corresponding controller during a grid fault. Dynamic requirements based on the step response have been defined in [62] and later in [51], which is now tested during the certifi-

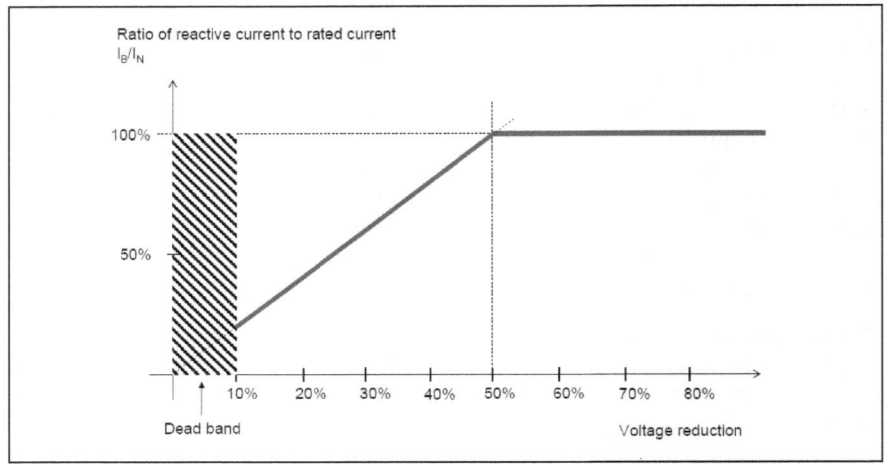

Fig. 6.2   Voltage control requirement for wind turbines in the case of sudden changes of grid voltage Source: E.ON 2003 [62].

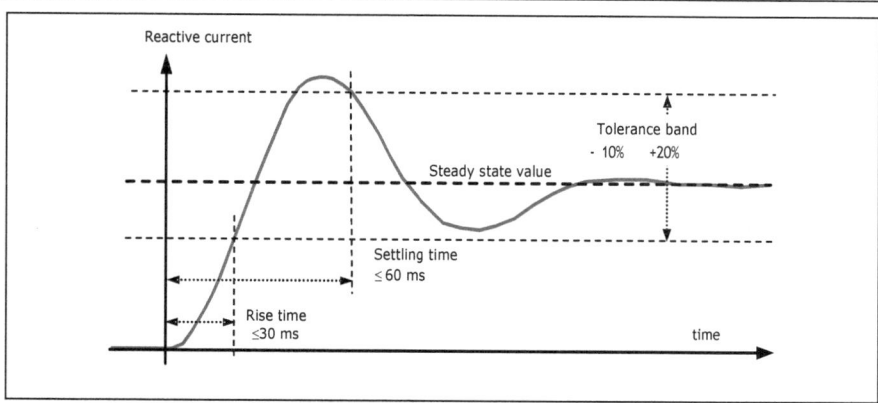

Fig. 6.3   Dynamic voltage control requirement for wind turbines in the case of sudden changes of grid voltage [51].

cation procedure in Germany. The rise time required is about 30 ms and the settling time 60 ms (see Fig. 6.3).

A definition based on reactive current (instead of reactive power) is common for describing the voltage control requirements for wind turbines during grid faults [51],[67]. In contrast to conventional synchronous generators, wind turbines allow a very fast simultaneous electronic control of both reactive and active currents. However, the power electronic components are sensitive to current limit violations. A temporary increase of reactive current at the expense of active currents is possible. But this can be done only for a limited time otherwise the generator could accelerate beyond the allowable limits.

An additional reason for a fast reactive current in-feed is the need for a sufficiently high fault current to trigger the grid protection. This current does not have to be very high if the fault at the same time leads to drops in the voltage level.

6.2.3 Summary of the Requirements

Two main tasks of the reactive power contribution of a power station can be identified:

(1) slow response to grid load flow and topology changes

(2) fast response during grid faults to mitigate the effect of fast voltage changes

By responding to load flow changes the speed of response should be limited also in order to avoid unnecessary control action. This control loop should not respond to the fast voltage changes.

The fast control by contrast has to respond mainly to fast changes in grid voltage. It should not alter the longer term settings of the slow controller but should modify the reactive current reference only following a voltage change.

Table 6.1 gives a summary of the requirements of slow and fast Var control loops for wind turbines:

Table 6.1   Comparison of slow and fast reactive power control requirements of power stations

Slow (outer) control loop	Fast (inner) control loop
Disturbance response	
Contribute to grid load flow and voltage control	Controlling voltage in the case of sudden voltage changes
Required response time: 7 s .. 60 s	Required response time: 20 ms ..30 ms
Should not respond to fast voltage changes to avoid unnecessary control actions	Must react to fast voltage changes to stabilize voltage level
Response to slow voltage changes as result of load flow changes	Response to fast changes of the voltage as reaction to grid faults and load rejection
Response to reference change	
Response to external reference by DSO or TSO	Response to local voltage change
Set long term reactive power reference	Additional reactive current contribution as reaction to voltage change
Within long term capability of power station	Use short-term overloading capability
Control at wind plant level	Control at wind turbine level
Control Characteristic	
Within the stationary capability limits of the wind turbines	Use of short term overloading capabilities of the wind turbines
Predominantly integral control	Predominantly proportional and/or differential control

6.3 Reactive Current Contribution of Synchronous Generators

Synchronous generators directly coupled to the grid are the most important source of electrical power in almost all grids. The concept of operating this generator at the grid is proven and well understood. Besides the gird requirements as summarized in table 6.1, the reactive power contribution of synchronous generators will therefore be taken as a reference for the design of a wind plant reactive power control.

For grid studies the high voltage connection to the grid is the relevant point to look at. Therefore, the main transformer of the unit has to be considered in the analysis of the synchronous generator and the wind plant response. The consumer oriented sign convention is applied for all equations and diagrams where consumed active power and inductive reactive power are considered to be positive.

6.3.1 Detailed Synchronous Generator Model

The common representation of the synchronous generator during steady state operating conditions is a voltage source behind and impedance (x_d, x_q). The "internal" generator voltage sources are often referred to as *electromagnetic force (emf)*. In case of a grid fault, a dynamic representation is necessary. Changes in grid voltage lead to temporary changes of the flux path in the generator ([63], p 135). Additional parameters describing transient and subtransient conditions are commonly used [63], [70].

During *subtransient mode*, the stator flux is forced into paths outside the rotor by currents induced in the field and damper winding. This effect is represented by the inductances x_d'' and x_q''. The subtransient currents decay fast with typical short circuit time constants T_d'', $T_q'' = 20$ ms .. 50 ms.

During *transient mode*, the stator flux is forced into paths outside the field winding by currents induced in the field winding. This effect is represented by the inductance x_d' and x_q'. The associated time constant T_d' has a typical value of 1 s. Transient and subtransient mode are associated with corresponding transient and subtransient voltages (v', v''). For a detailed analysis, both d and q-axis are considered (see Fig. 6.4).

In *steady state* the effective driving voltage is derived from the field voltage v_{fd} supplied by the exciter system.

The three states of synchronous machines are described by the following equations:

Steady state:

$$v_{sd} = r_s i_{sd} + x_q i_{sq} \tag{6.1}$$

$$v_{sq} = r_s i_{sd} + x_d i_{sd} + \frac{x_{md}}{r_{fd}} v_{fd} \tag{6.2}$$

$$\underline{v}_s = \left(r_s + j x_d\right)\underline{i}_s + \underline{v}_p \tag{6.3}$$

with

v_{sd}, v_{sq}; $\underline{v}_s = v_{sd} + j v_{sq}$	stator terminal voltage
r_s	stator resistance
i_{sd}, i_{sq}; $\underline{i}_s = i_{sd} + j i_{sq}$	stator current
x_d, x_q	synchronous reactances
x_{md}	mutual reactance in the d-axis
v_{fd}	field winding voltage
r_{fd}	field winding resistance
$v_p = j \frac{x_{md}}{r_{fd}} v_{fd}$	(inner) rotor pole voltage

Transient state:

$$v_{sd} = r_s i_{sd} + x_q' i_{sq} - \omega_0 k_{fq} \psi_{fq} \tag{6.4}$$

$$v_{sq} = r_s i_{sq} + x_d' i_{sd} + \omega_0 k_{fd} \psi_{fd} \tag{6.5}$$

$$\underline{v}_s = \left(r + j x_d'\right)\underline{i}_s + \left(x_q' - x_d'\right)i_{sq} + \underline{v}' \tag{6.6}$$

$$\underline{v}' = v_d' + j v_q' = -\omega_0 k_{fq} \psi_{fq} + j \omega_0 k_{fd} \psi_{fd} \tag{6.7}$$

with

x_d', x_q'	transient reactances
k_{fd}, k_{fq}	field winding coupling factors
ψ_{fd}, ψ_{fq}	field flux linkages[7]
$\underline{v}' = v_d' + j v_q'$	transient voltage

---

[7] The index fd is used in equivalence to the d-axis, fd represents a second equivalent damper winding in the q-axis.

Subtransient state:

$$v_{sd} = r_s i_{sd} + x_q'' i_{sq} - \omega_0 \left(k_{fq} \psi_{fq} + k_{Dq} \psi_{Dq} \right)$$ (6.8)

$$v_{sq} = r_s i_{sq} + x_d'' i_{sd} + \omega_0 \left(k_{fd} \psi_{fd} + k_{Dd} \psi_{Dd} \right)$$ (6.9)

$$\underline{v}_s = \left(r_s + j x_d'' \right) \underline{i}_s + \left(x_q'' - x_d'' \right) i_{sq} + \underline{v}''$$ (6.10)

$$\underline{v}'' = -\omega_0 \left(k_{fq} \psi_{fq} + k_{Dq} \psi_{Dq} \right) + j \omega_0 \left(k_{fd} \psi_{fd} + k_{Dd} \psi_{Dd} \right)$$ (6.11)

with

x_d'', x_q'' — subtransient reactances

k_{Dd}, k_{Dq} — damper winding coupling factors

ψ_{Dd}, ψ_{Dq} — damper winding flux linkages

$\underline{v}'' = v_d'' + j v_q''$ — subtransient voltages

The rotor flux linkages and the emfs are assumed to remain constant for the time of grid faults at the pre-fault level. This is commonly referred to as *law of constant flux linkages*. It basically states that the energy stored in the magnetic field cannot change instantaneously. However, in reality following disturbances like short circuits, a transition from subtransient through transient to steady state model take place. For more accurate calculations the change of rotor field and damper fluxes can be described by differential equations as shown above, see also [70].

The response of synchronous generators to set point changes is slow compared to converter based systems, they will only lead to slow changes of reactive power[8]. The control of the exciter system is not modeled for the short circuit analysis, the field voltage is therefore assumed to stay constant during the fault. The impact of the field voltage control is analyzed later during dynamic simulation (see Fig. 6.11).

---

[8] Typical values are up to 0.3 s for static exciters and up to 1 s for rotating exciters [63], p 25.

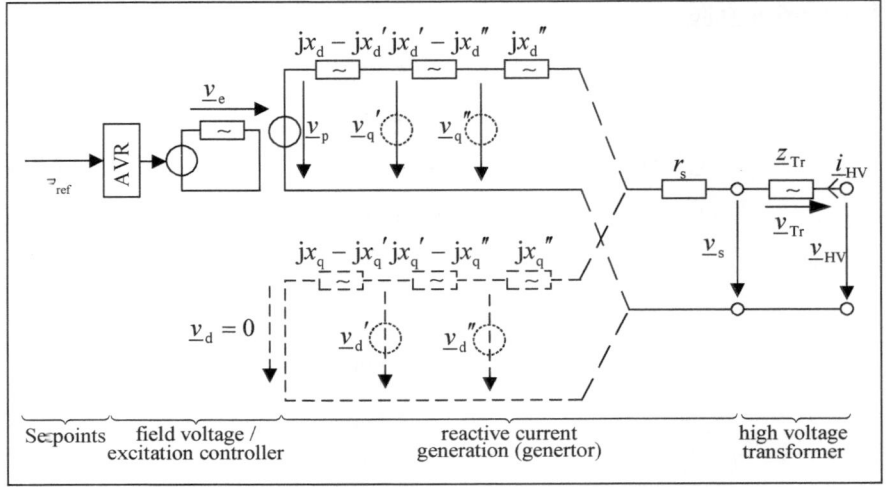

Fig. 6.4 Simplified model representation of synchronous generator and grid transformer.

If the generator is assumed to run on no load before the grid fault, the terminal voltage equals the emf,

$$\underline{v}_p = \underline{v}'_q = \underline{v}''_q = \underline{v}_s = \underline{v}_{HV} - \underline{v}_{Tr} \tag{6.12}$$

If the generator is operating under load, the emfs for each mode have to be calculated using the grid voltage, the generator current and the inductances. Fig. 6.5 shows a simplified representation of a synchronous machine that will be used for calculating the initial emfs, neglecting saliency, i.e. setting $x'_d = x'_q$, $x''_d = x''_c$. The armature resistance is small and can be ignored.

For a pre-specified active and reactive power as well as voltage at the HV terminals, we have for the current $\underline{i}_{HV} = (p_{HV} - jq_{HV})/\underline{u}^*_{HV}$. The initial emfs can be calculated as:

$$\underline{v}_p = \underline{v}_{HV} + \underline{i}_{HV}\left(jx_d + \underline{z}_{Tr}\right) \tag{6.13}$$

$$\underline{v}' = \underline{v}_{HV} + \underline{i}_{HV}\left(jx'_d + \underline{z}_{Tr}\right) \tag{6.14}$$

$$\underline{v}'' = \underline{v}_{HV} + \underline{i}_{HV}\left(jx''_d + \underline{z}_{Tr}\right) \tag{6.15}$$

6.1.2 Simplified Model for Current Calculation during Grid Faults

An approximation of the currents of a synchronous generator during a grid fault can be achieved using the equivalent circuit presented in Fig. 6.5. The grid fault representation is based on a load $\underline{z}_{load}$ and a fault impedance $\underline{z}_{fault}$ that is varied from low to very high values to emulate different (electrical) distances of the generator from the fault location.

For a sample calculation, the pre-fault voltage and power are set to $v_{HV} =$ 1 p.u., $p_{HV} = 1$ p.u., $q_{HV} = 0$ p.u.. The load z_{load} can then be calculated as

$$\underline{z}_{load} = \underline{v}_{HV} \cdot \underline{v}_{HV}^* / \left(p_{HV} - jq_{HV} \right) \tag{6.16}$$

The fault impedance is assumed to have a r/x ratio of 1:5. As a result, the r/x-ratio of the grid model changes as the fault comes (electrically) closer to the generator. A large wound rotor generator with identical parameters for d- and q-axis is assumed with $x_d = x_q = 2$ p.u., $x_d' = x_q' = 0.4$ p.u., $x_d'' = x_q'' = 0.25$ p.u. and a transformer impedance of $x_{Tr} = 0.12$ p.u. and $r_{Tr} = 0.006$ p.u.

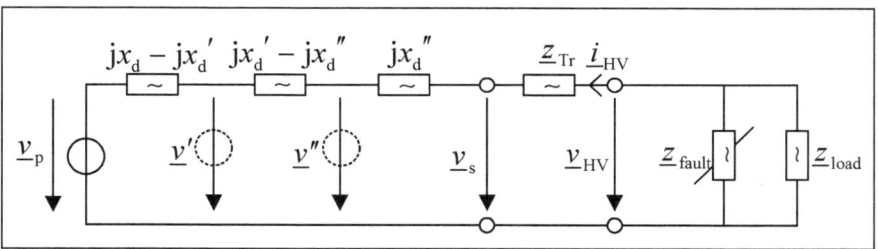

Fig. 6.5   Simplified model of synchronous generator and grid fault for the estimation of reactive currents during faults. The variation of the value of z_{fault} emulates the electrical distance of the fault from the generator.

Fig. 6.5 shows the model of the synchronous generator together with a load and a fault impedance for the estimation of reactive currents during faults. The variation of the value of z_{fault} emulates the electrical distance of the fault from the generator. The resulting grid impedance is varied using

$$\underline{z}_{grid} = k \cdot \underline{z}_{fault} \parallel \underline{z}_{load} = \frac{k \cdot \underline{z}_{fault} \cdot \underline{z}_{load}}{k \cdot \underline{z}_{fault} + \underline{z}_{load}} \tag{6.17}$$

Assuming constant emf during the fault, the short circuit currents then are

$$\underline{i}_k = \frac{\underline{v}_s}{jx_d + jx_{Tr} + \underline{z}_{grid}} \tag{6.18}$$

$$\underline{i}_k' = \frac{\underline{v}'}{jx_d' + jx_{Tr} + \underline{z}_{grid}} \tag{6.19}$$

$$\underline{i}_k'' = \frac{\underline{v}''}{jx_d'' + jx_{Tr} + \underline{z}_{grid}} \tag{6.20}$$

The steady state current is calculated for comparison only. The value is not relevant for short circuit consideration.

The resulting voltages at the HV terminals of the transformer and at the synchronous generator at MV level can then be calculated for the steady state, transient and subtransient currents :

$$\underline{v}_{\text{HV,fault}} = \underline{i}_k \underline{z}_{\text{grid}} \qquad \underline{v}_{\text{MV,fault}} = \underline{i}_k \left(\underline{z}_{\text{grid}} + \underline{z}_{\text{TrHv}} \right) \tag{6.21}$$

$$\underline{v}'_{\text{HV,fault}} = \underline{i}'_k \underline{z}_{\text{grid}} \qquad \underline{v}'_{\text{MV,fault}} = \underline{i}'_k \left(\underline{z}_{\text{grid}} + \underline{z}_{\text{TrHv}} \right) \tag{6.22}$$

$$\underline{v}''_{\text{HV,fault}} = \underline{i}_k'' \underline{z}_{\text{grid}} \qquad \underline{v}''_{\text{MV,fault}} = \underline{i}_k'' \left(\underline{z}_{\text{grid}} + \underline{z}_{\text{TrHv}} \right) \tag{6.23}$$

6.3 3 Static Calculation of Synchronous Generator Reactive Current Gain

The relation between voltage changes due to the faults and reactive current contribution can be described as gain, see (6.24) for the high voltage and medium voltage (stator) terminals:

$$k_{\text{iQ, HV}} = \frac{\Delta i_{\text{QHV}}}{\Delta v_{\text{HV}}} \quad \text{and} \quad k_{\text{iQ, MV}} = \frac{\Delta i_{\text{QMV}}}{\Delta v_{\text{MV}}} \tag{6.24}$$

with $\Delta v_{\text{HV}} = \left| \underline{v}_{\text{HV}}(0) \right| - \left| \underline{v}_{\text{HV}}(\text{fault}) \right|$ and $\Delta i_{\text{Q,HV}} = -\{ i_{\text{Q,HV}}(0) - i_{\text{Q,HV}}(\text{fault}) \}$.
where the reactive current i_{Q} is defined in general by

$$i_{\text{Q}} = -\,\text{Im}\left\{ \underline{i}^{\angle v} \right\} \tag{6.25}$$

The sign $\angle v$ indicates that the current is considered in a coordinate system where the voltage is laying along the real axis. The reactive current contribution can be defined similarly also for the medium voltage side.

The dependency of the reactive short circuit current as function of the remaining HV grid voltage is shown in Fig. 6.6 for the steady state, transient and subtransient reactive current. The reactive current increases in close to linear fashion with the voltage decrease. The impact of active power level on the reactive current gain is small and can be neglected.

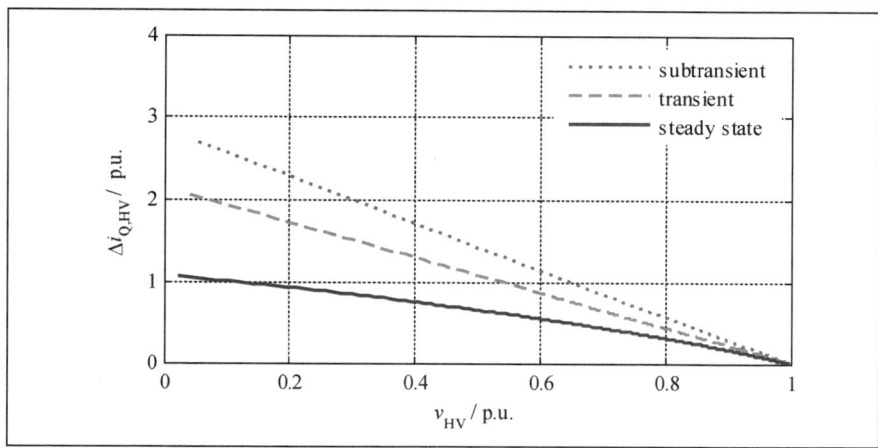

Fig. 6.6 Change of HV subtransient, transient and steady state HV reactive currents of a synchronous generator during grid faults as function of voltage change.

Fig. 6.7 shows the resulting gain k_{iQ} calculated according to (6.24), describing the ratio of reactive current increase due to voltage decrease.

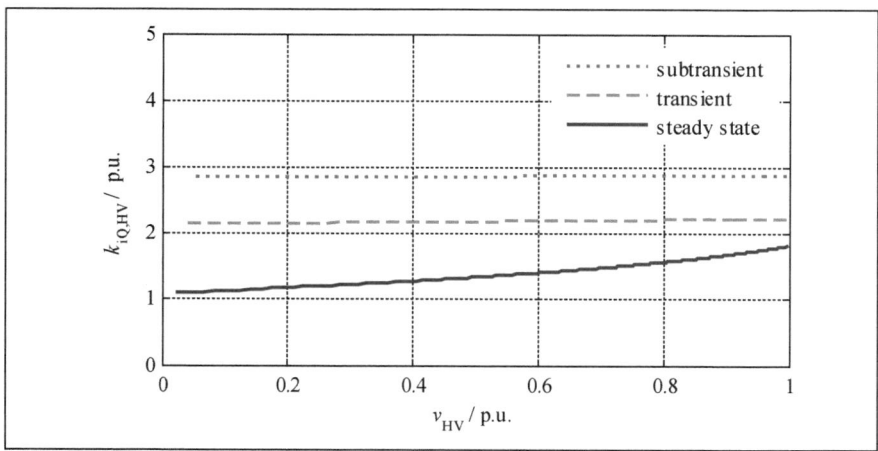

Fig. 6.7 Synchronous generator HV subtransient, transient and steady state reactive current gain $k_{iQ,HV} = \Delta i_{Q,HV}/\Delta v_{HV}$ as function of voltage change.

The equivalent diagrams of reactive current change and reactive current gain at the medium voltage terminals are shown in Fig. 6.8 and Fig. 6.9.

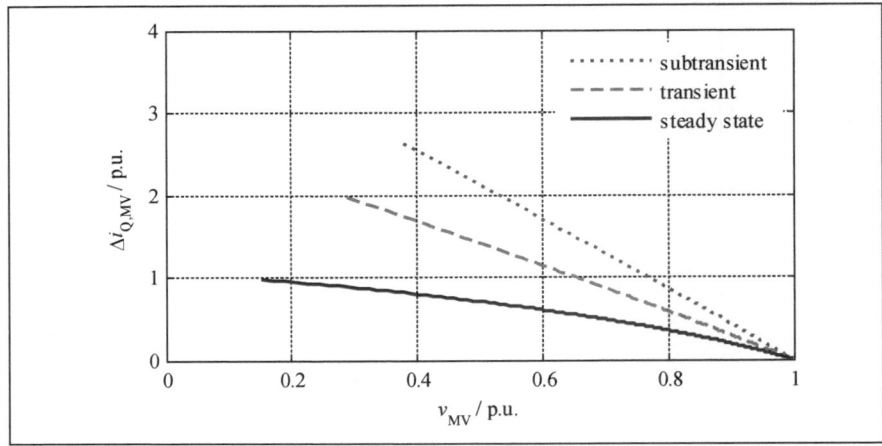

Fig. 6.8   Change of MV subtransient, transient and steady state reactive currents of a
synchronous generator during grid faults at the high voltage terminals as function of
voltage change.

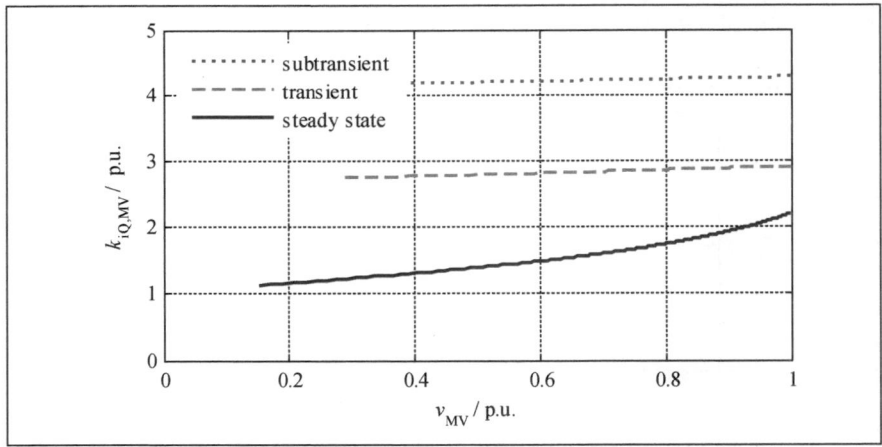

Fig. 6.9   Synchronous generator MV subtransient, transient and steady state reactive current
gain $k_{iQ,MV} = \Delta i_{Q,MV}/\Delta v_{MV}$ as function of voltage change.

As shown, the reactive current contribution of a synchronous generator at
the high voltage terminals during grid faults can be approximated by a linear gain
that describes the ratio of average reactive current supplied during fault and voltage change. In this example, the maximum value of the gain is between 2.2 p.u.
for the transient currents and 2.9 p.u. for the subtransient currents. The impact of
the subtransient current is mainly relevant at the beginning of the fault. The sub-

transient time constant for electrically close faults is small (usually between 50 ms for small generators and 20 ms for very large generators). For electrically distant faults, the effective time constant is larger , see also Fig. 6.11 in the next section.

6.3.4 Dynamic Simulation of Reactive Current Gain of Synchronous Generators

The results of the static calculations of the reactive current contribution of synchronous generators (as shown in Fig. 6.6 - Fig. 6.9) are now compared to a dynamic simulation to validate the results.

A power station using a synchronous generator connected to the high voltage grid as shown in Fig. 6.10 is used for the simulation. A detailed simulation of a 50 MW (55 MVA) synchronous generator using a 7th order model is used to compare the calculated values to the results of the simulation. The model uses an excitation control based on [71], and a model of the mechanical and governor control based on [34]. The simulation software used is Matlab/Simulink with the SimPowerSystem Toolbox [72].

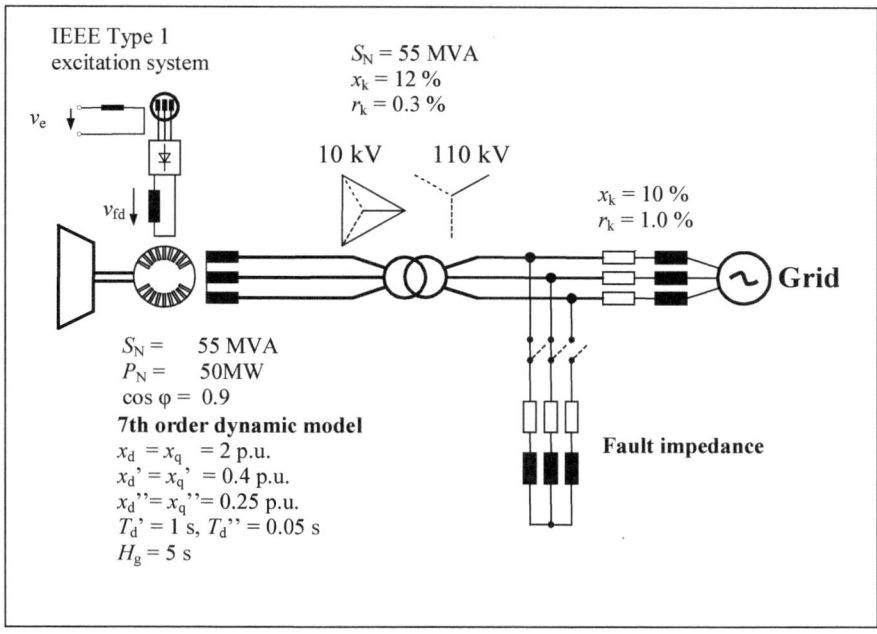

Fig. 6.10 Configuration of 50 MW synchronous generator.

Impact of Field Voltage Control of the Detailed Model

Reactive power control of the synchronous generators under normal operating conditions is performed by controlling the excitation voltage of the generator (see also Fig. 6.4). The excitation voltage is controlled by the excitation controller that receives setpoints from the system operator. Usually the generator voltage is controlled, the excitation control is then referred to as automatic voltage regulator (AVR).

Fig. 6.11 shows the voltage and reactive current at the high voltage terminals both for active excitation control using a voltage regulator (AVR) and for constant field voltage during the fault as it had been assumed in the static calculation of the short circuit currents in (6.20). A maximum field voltage of 6 p.u. is assumed in the simulation.

A constant field voltage during the fault leads to a slightly faster decrease of the reactive current than activated field voltage control. The impact of the field voltage control is limited, and the deviation is less than 2.5% of the reactive current at the end of the grid fault. As an average over the fault, the deviation is less than 1%.

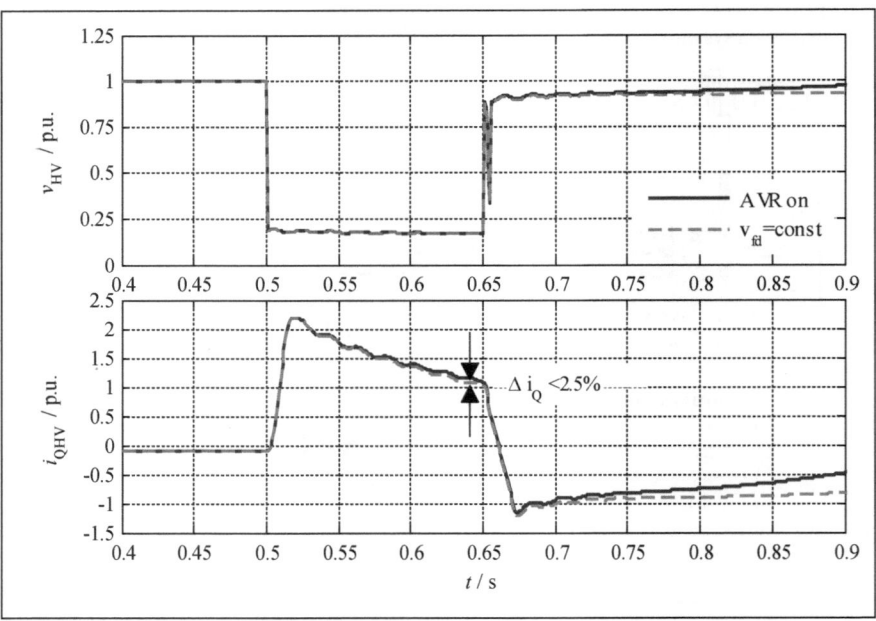

Fig 6.11 Impact of field voltage control on reactive current during a grid fault (The reactive current shown using producer oriented sign system).

Comparison of Static and Dynamic Reactive Current Gain Calculation

The dynamic reactive current gain during the grid fault of the dynamic simulation is shown in Fig. 6.12. How fast subtransient and transient currents decay also depends on the electrical distance of a fault. For a distant fault with smaller voltage sag at the generator terminals, the currents decay slower. Fig. 6.12 therefore shows the reactive current gain both for a close and a distant fault with a remaining voltage at the high voltage terminals of 0.2 p.u. for the close fault and 0.85 p.u. for the distant fault.

As can be seen, the highest current gain is reached shortly after the beginning of the fault. The maximum gain from the dynamic calculations is almost equivalent to the result of the static calculation in Fig. 6.9 for the contribution of the subtransient component x_d''. Due to the short subtransient time constant T_d'' of 0.05 s this value decreases fast.

The transient time constant of T_d' (1 s) is long compared to typical grid faults (0.1 s .. 0.5 s), therefore the transient current component changes only slowly during the fault. As comparison, Fig. 6.12 also shows:

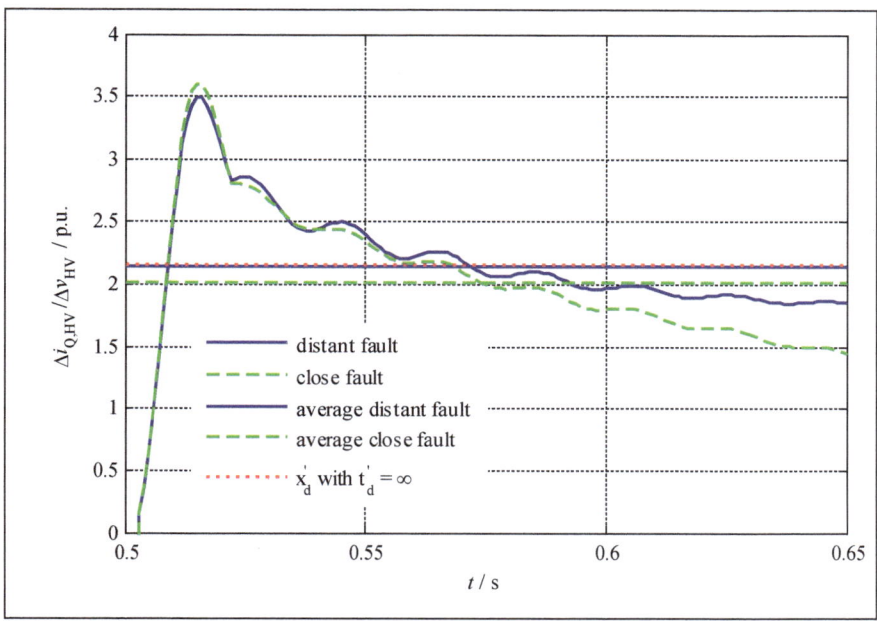

Fig. 6.12 Reactive current gain $k_{iQHV} = \Delta i_{QHV}/\Delta v_{HV}$ for a detailed simulation of a 50 MW synchronous generator during a fault at the high voltage terminals of the transformer.

■   an average gain over the entire fault for the close and the distant fault, and

■   an equivalent gain calculated based on the fixed transient voltage behind the
    reactance x_d' assuming a time constant of $T_d' = \infty$.

Equivalent Reactive Current Gain for Wind Plants

Based on the analysis, a *constant gain* of 1.9 p.u. .. 2.2 p.u. at the high voltage
terminals can be assumed as a good reference for the design of a wind plant controller if the same response as of a synchronous generator is demanded. This current should be *in addition* to the pre-fault reactive current.

6.4 Wind Plant Implementation of Reactive Power Control

In comparison to a synchronous generator where plant control and generator are
located very close together, a wind plant is may stretch over a large area. Therefore, communication delays between measurement, control and actuators have to
be considered. The fast response needed during grid faults requires also a short
distance between measurement and control. While centralized slow control loops
allow a precise control at the point of common coupling, a fast control needs to
be implemented at the wind turbine level. This is also reflected in grid codes [68]
where control requirements of turbines during grid faults are defined at the turbine terminals.

The following core requirements were deduced in the previous section:

1)   slow control loop at plant level

2)   fast control loop at turbine level during fast voltage changes

3)   an additional proportional reactive current contribution during faults

6.4.1   Wind Plant Controller Design

The plant level set points and control reactions are usually defined with respect to
the point of common coupling (PCC). The task of the wind plant controller is to
control the reference signals of the individual wind turbines and possible additional active or passive reactive power compensation equipment, so the control
requirements at the PCC are met.

The basic structure of a wind plant reactive power control with the operation
modes as described in section 6.2.1 is given in Fig. 6.13. Different inputs for the
set points are possible depending on the grid code or grid operator needs.

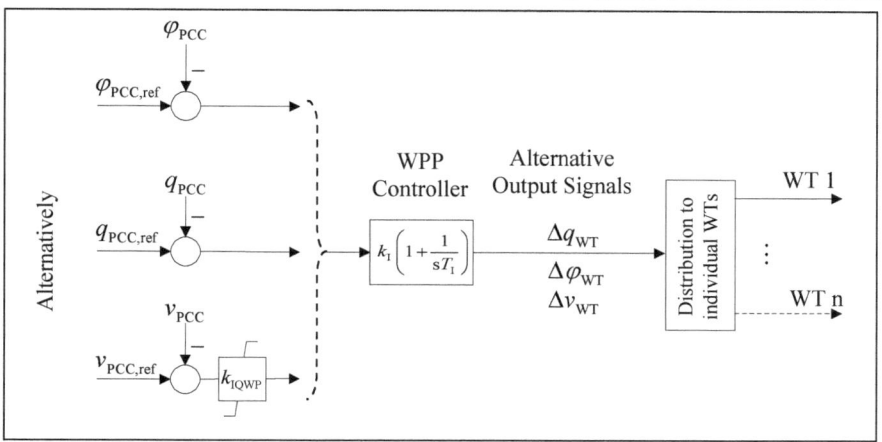

Fig. 6.13 Fundamental structure of wind power plant voltage-Var controller.

A PI control is used to generate output signals. Common implementations use reactive power reference, power factor reference or voltage reference. Finally, the output signals are distributed to the individual wind turbines (WT). This block also represents the time delay resulting from the transport of the signal from a centralized controller to the turbines.

Voltage Static

A special form of voltage control is a voltage static (also referred to as Q(V) control) as it is required for example in the UK grid code requirement [24], but also in [69] and [67]. The UK grid code defines reactive power demand as function of a specified voltage tolerance around rated voltage. For a target operation range of ± 5% and a reactive wind plant rating of ± 0.33 p.u., a gain can be defined as

$$k_Q = \frac{\Delta q}{\Delta v} = \frac{q_{max} - q_{min}}{v_{max} - v_{min}} = \frac{0.33 \text{ p.u.} - (-0.33 \text{ p.u.})}{1.05 \text{ p.u.} - 0.95 \text{ p.u.}} = \frac{0.66 \text{ p.u.}}{0.1 \text{ p.u.}} = 6.6 \text{ p.u.} \quad (6.26)$$

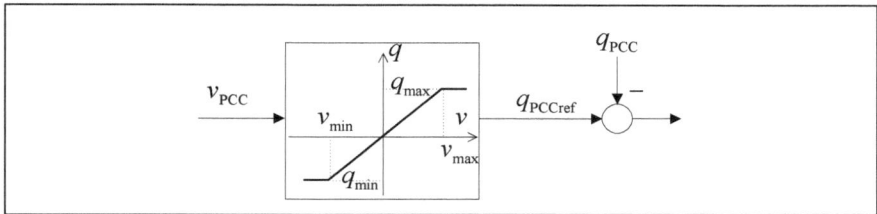

Fig. 6.14 Wind power plant Var control using voltage static.

The maximum reactive power will be demanded if the voltage limits are reached. An example of the implementation of a reactive power versus voltage (QV) characteristic is given in Fig. 6.14.

Implementation

An implementation of the wind plant control is shown in Fig. 6.15 and in more detail in Fig. 6.16. The control structure consists of a set point generation block, the wind plant control and the communication delay. The set point generation allows the selection of different control values depending on the requirements of the grid code and the turbine site. The input values are converted into a reactive power reference. The voltage static is converted to a structure that allows the modification of the voltage reference using an external reference.

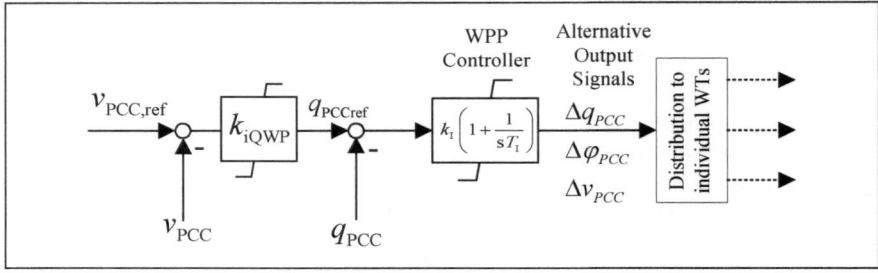

Fig. 6.15 Implementation of wind plant Var control using voltage static.

The plant control is represented by a PI-control loop for the reactive power including limitation blocks. The wind plant reference values need to be limited to the wind plant operating range. The turbine reference limit should reflect the turbine control limitations. All measured values of the wind plant controller (v_{PCC}, q_{PCC}.) should be subject to some filtering to avoid undesired control actions.

It is assumed that all turbines receive the same reference signal, in reality optimizations may take place that provide individual turbines with different reference signals. The communication delay represents the signal delays resulting from the controller cycle time, transport delays of the communication between plant controller and wind plants and optional filtering in the wind turbine control.

If the voltage is outside a predefined normal operating range under fault conditions, the integral part of the PI control should to be reset or frozen in order to avoid unintended plant control activity during the fault and a delayed recovery following grid faults (see Fig. 6.16).

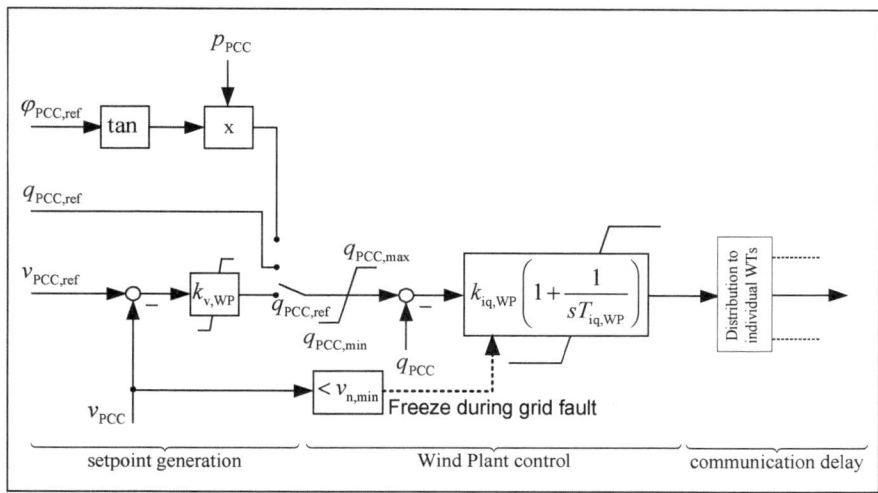

Fig. 6.16 Implementation of wind plant Var control.

6.4.2 Control Structure using Voltage Reference at Turbine Level

As described, a reactive power or reactive current control for wind plants will need two levels of control:

(1) a plant controller receiving set points from the grid operator and calculating a voltage reference. The plant controller is the outer, slow control loop that has to respond to load flow changes.

(2) a system generating reactive currents as function of grid voltage and wind plant controller voltage reference. This is the fast local control loop that needs to react fast on voltage changes.

The reactive current demand for a wind plant that has an fast reactive current response to voltage changes comparable to synchronous generators connected to the high voltage grid can be described by a voltage source behind an inductance with an impedance defined by the desired reactive power / voltage gain k_{iQ} with

$$jx_{WP} = j\frac{1}{k_{iQ}} \tag{6.27}$$

In a real wind plant implementation, a fast control always needs to be at turbine level. This is faster, avoiding measurement and communication delays between the plant controller and the turbine and is more reliable; it does not depend on fast signals from a source outside of the turbine. A corresponding representation at turbine level is shown in Fig. 6.17. The corresponding response of the reactive current as function of remaining voltage is shown in Fig. 6.18 (using producer oriented sign system for compatibility with [62]).

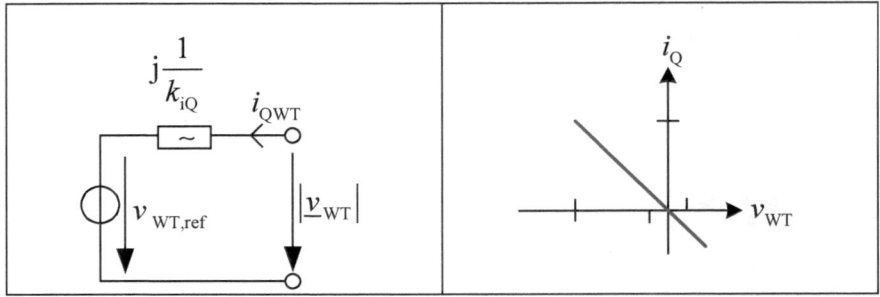

Fig. 6.17 Equivalent representation of reactive current generation reference for a wind turbine.

Fig. 6.18 Resulting reactive current contribution of a wind plant during grid faults.

A difference between synchronous generators and wind plants with converter based generator systems (DFG, FSC) is that the active power can be controlled independently from the reactive power and independently from mechanical torque or speed of the generator. For a simplified representation, the active current can be represented as a current source. The resulting structure is shown in Fig. 6.19.

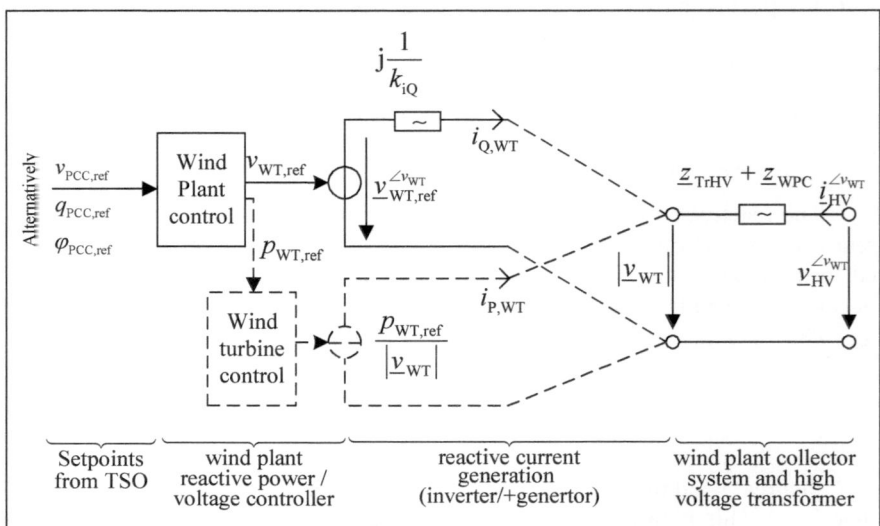

Fig. 6.19 Simplified wind plant reactive power control model.

A comparison to the structure of a synchronous power plant (see Fig. 6.4) shows some similarities, in the case of the wind turbine, a voltage reference signal ($v_{WT,ref}$) is used to control the reactive power output, in the case of the synchronous generator, the field voltage (i.e. not the reference) controls the reactive power output. In both cases, the reactive current during grid faults is nearly proportional to the voltage change (within the current limits of the turbine).

Wind Turbine Controller Design

A reactive current control law for a wind turbine can then be described as

$$i_{QWT,ref} = k_{iQ}\left(v_{WT,ref} - v_{WT}\right) \tag{6.28}$$

with $v_{WT,ref}$ as the reference voltage calculated by the plant controller, v_{WT} voltage at the turbine terminals and k_{iQ} as the demanded proportional gain. This is shown in Fig. 6.20. In Fig. 6.21, the reactive current is shown as function of voltage (using producer oriented sign system for compatibility with [62]). By changing the voltage reference, the red curve is moved along the x-axis.

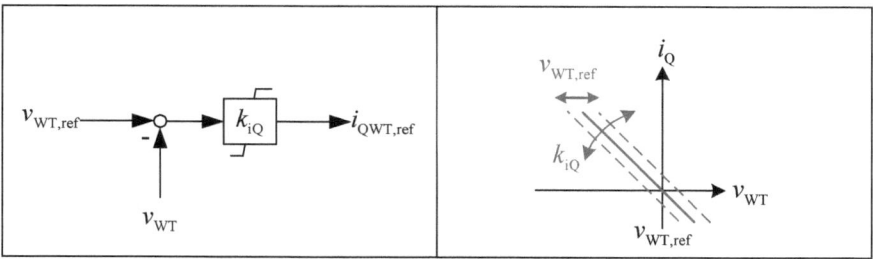

Fig. 6.20 Turbine level reactive current control law.

Fig. 6.21 Demanded dynamic reactive current contribution of a wind turbine.

Compensation of the reactive demand of the connection line.

As described before, the fast control will always be at turbine level. In case a proportional control of the reactive current as function of the HV voltage is desired, the impedance between turbines and wind plant terminals is relevant because it leads to a voltage change between turbine and PCC. As shown in Fig. 6.22, possible impedances are the high voltage transformer impedance z_{TrHV} and cable impedances in the medium voltage level resulting from the wind plant collectors system described by $\underline{z}_{MV}$. In case the turbine control acts at low voltage level, also the medium voltage transformer needs to be considered in this parameter.

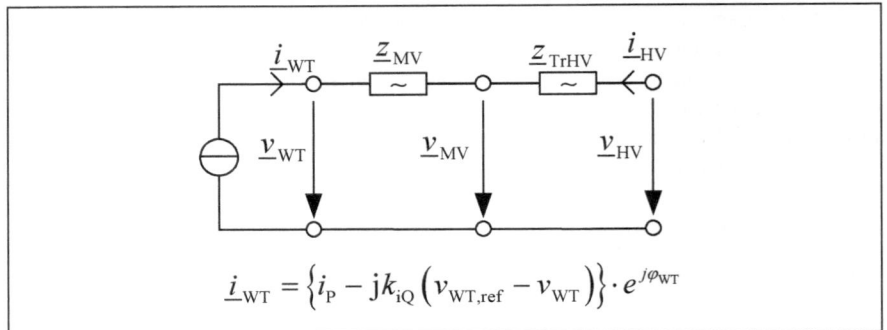

$$\underline{i}_{\mathrm{WT}} = \left\{i_{\mathrm{P}} - jk_{\mathrm{iQ}}\left(v_{\mathrm{WT,ref}} - v_{\mathrm{WT}}\right)\right\} \cdot e^{j\varphi_{\mathrm{WT}}}$$

Fig. 6.22 Impedances of wind plant.

One option would be to define a higher reactive current gain at turbine level in order to achieve a specified gain at the HV terminals as proposed for example in [73], p 95. A different approach proposed here is to include the plant impedances into the control law:

$$i_{\mathrm{QWT,ref}} = k_{\mathrm{iQ}}\left(v_{\mathrm{WT,ref}} - \left|\underline{v}_{\mathrm{WT}} + \underline{i}_{\mathrm{WT}}\left(\underline{z}_{\mathrm{MV}} + \underline{z}_{\mathrm{TrHV}}\right)\right|\right) \tag{6.29}$$

with $\underline{z}_{\mathrm{MV}}$ as the impedance of medium voltage transformer and wind plant collector system and $\underline{z}_{\mathrm{TrHV}}$ as the impedance of the high voltage transformer. The use of a voltage droop calculation based on (6.29) effectively 'moves' the control point further into the grid.

In case several wind plants of synchronous generators feed in parallel onto a common bus bar, the compensation should not be higher than 85% of the impedance of $\underline{z}_{\mathrm{MV}} + \underline{z}_{\mathrm{TrHV}}$ to keep a sufficient 'electrical distance' to other generators ([63], p 24). For wind plants, the effective impedance between a wind plant and the high voltage terminals is most likely different for every wind turbine. For the practical use, an average impedance $\underline{z}_{\mathrm{WP}}^{*}$ of the wind plant needs to be calculated.

The impact of the cable capacity on the reactive power could be included, but for an operation during grid faults this is usually not necessary because the accuracy requirements are reduced. For normal operating conditions at rated voltage with limited changes of the voltage within the normal operating range, the impact is usually rather limited.

The resulting control structure including voltage estimation is shown in Fig. 6.23. This solution can be useful in the case a voltage static with very short settling time is demanded at plant level (as required for example in [24]). In this case, the gain of the voltage static at plant level and the local gain k_{iQ} at the turbine level can be set to the same value. The impact of transformers and feeders in

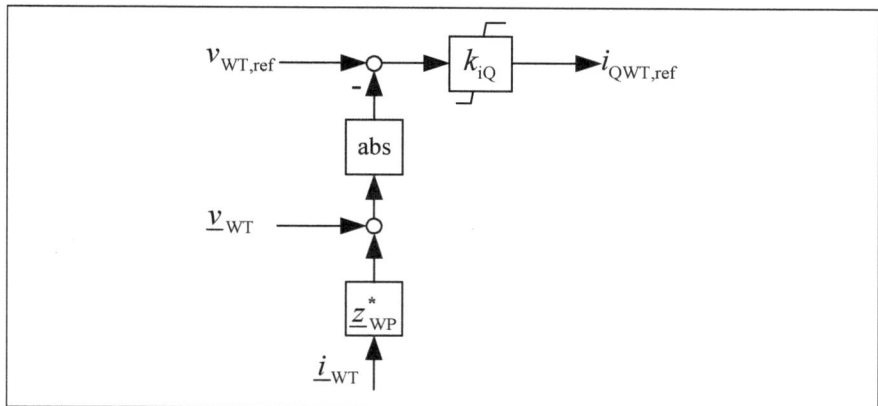

Fig. 6.23 Wind turbine reactive current control including droop compensation at turbine level.

the wind plant is compensated by $\underline{z}^*_{WP}$. An alternative if a compensation according to Fig. 6.23 is not used would be to specify a higher gain k_{iQ} at turbine level.

In both cases, the necessary control action by the plant controller can by minimized. The turbines will already provide most of the required control action, the plant controller is only required for a 'fine-tuning' with very limited changes to the turbine reference. This reduces the demands on the communication speed between wind plant and turbine while still providing sufficient control quality.

Var Control scheme by setting the voltage reference of individual wind turbines using the plant Var controller

The following simplified structure of wind plant and turbine controller was deduced (see Fig. 6.24). Communication between the plant controller and the turbine is based on voltage reference.

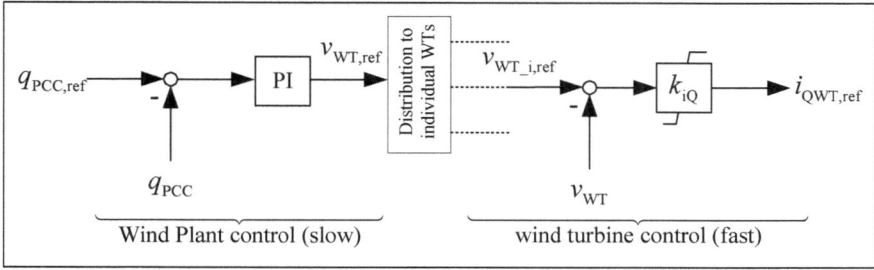

Fig. 6.24 Structure of wind plant Var control based on voltage reference.

In a practical implementation, it can be useful to modify the structure of Fig. 6.24 in a way that instead of a voltage reference v_{ref} a voltage difference Δv_{ref} is sent out by the plant controller. This has the advantage, that (a) in the case of a communication failure each wind turbine still has a defined reference and (b) it allows modifying the individual turbine output while keeping a single reference for all turbines, thus minimizing communication. In extension to Fig. 6.24 a voltage static and limitation functions at the plant controller are added in Fig. 6.25.

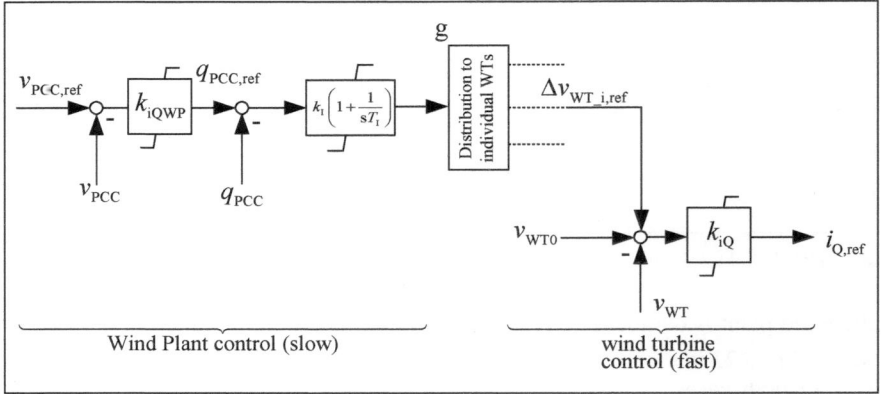

Fig. 6.25 Implementation of wind plant Var control based on voltage reference at turbine level.

6.4.3 Control Structure using Reactive Power or Reactive Current Reference at Turbine Level

A different implementation of the requirements deduced before:

(1) slow control loop at plant lever

(2) fast control loop at turbine level during fast voltage changes

(3) additional proportional reactive current contribution during faults

can be based not on voltage reference but on a reactive current reference, reactive power reference or phase angle reference from the plant controller. This allows the implementation of a control structure without deadband even for existing control topologies based for example on reactive power reference that so far often use a deadband in the reactive current control during grid faults.

A conceptual description of the control scheme that fulfills the above mentioned requirements is given in Fig. 6.26, part a). The resulting reactive current reference is the sum of a stationary component $i_{Q,ref0}$ (to control the voltage/reactive power changes that result from load flow changes) and a fast, dynamic component $\Delta i_{Q,ref}$ (as response to sudden voltage changes). By using a washout filter it can be assured that Δv_{VS} only responds to fast voltage changes and not to stationary changes of the voltage. The gain k describes the relationship

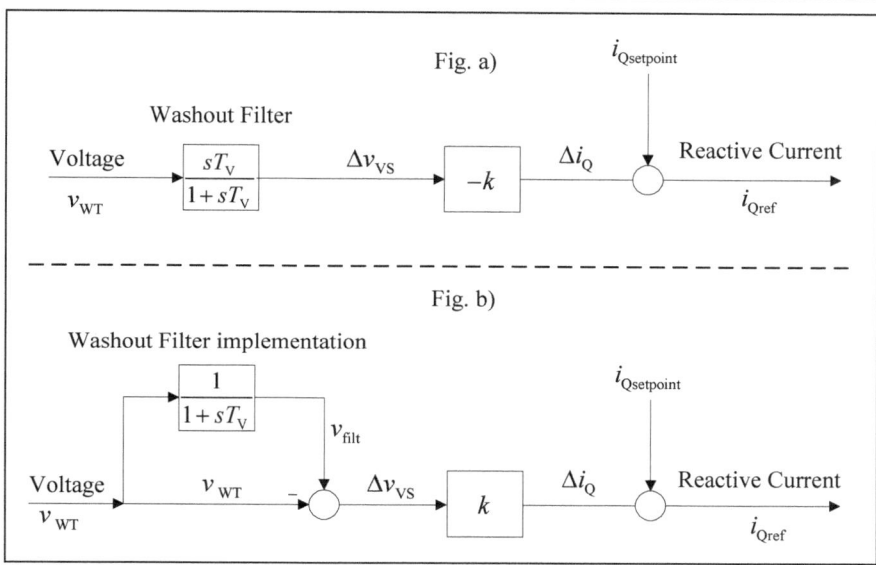

Fig. 6.26 Basic structure of continuous local voltage control at plant level with reactive current reference.

between the voltage change and the resulting reactive current change. A practical implementation of the washout filter is given in Fig. 6.26, part b).

Var Control scheme by setting the reactive current reference of individual wind turbines using the plant Var controller

An implementation including a simple plant controller is shown in Fig. 6.27. The resulting reactive current is the sum of a steady state component $i_{QWT,ref}$ for controlling load flow and a fast acting component Δi_Q that responds to voltage changes.

As in the case of the voltage reference, there is a slow loop (given by the plant controller) and a fast loop reacting to voltage changes. Since the voltage reference is now not given by the plant controller, the washout filter is used for creating a reference value that responds to fast voltage changes.

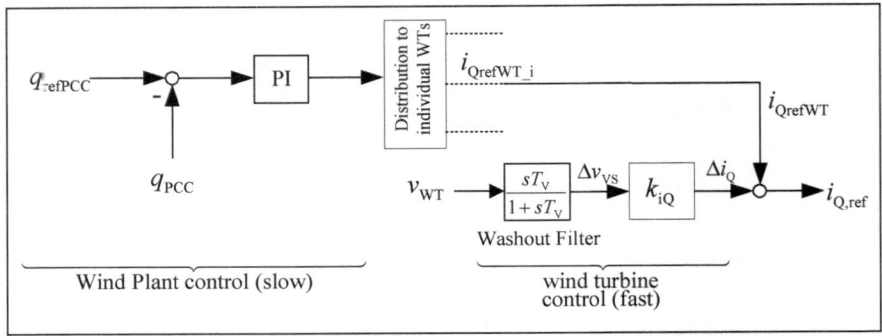

Fig. 6.27 Implementation of wind plant Var control based on reactive current reference.

It can be shown that the control structures based on Fig. 6.24 and Fig. 6.27 are equivalent with respect to the demanded dynamic control action. Based on the structure shown in Fig. 6.25, the value of v_{WT0} is replaced by a filtered voltage v_{filt} as shown in Fig. 6.27. The resulting structure corresponds to an implementation of a washout filter as shown Fig. 6.28.

In a second step, the summation point of the voltage reference from the plant controller is moved behind the gain k_{iQ} (see Fig. 6.28). This does not change the results, the voltage reference now has to be multiplied with the gain k_{iQ}. If this multiplication is moved into the plant controller, the structure is identical to the one shown in Fig. 6.27.

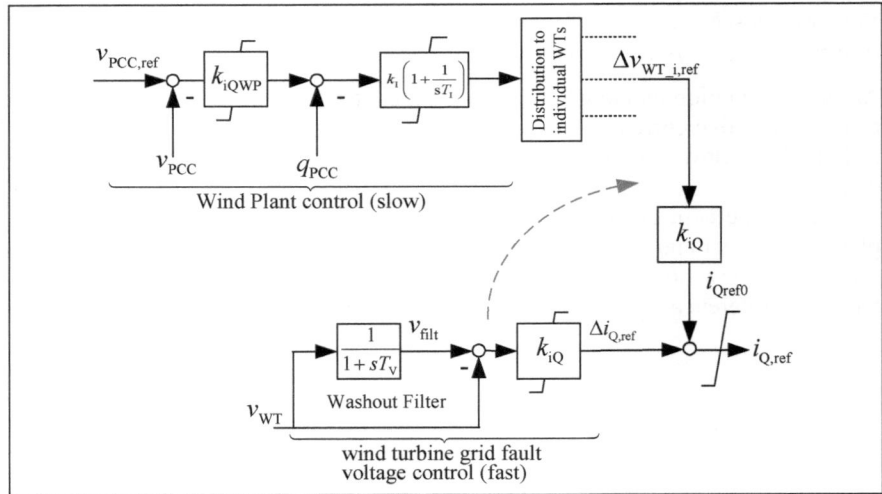

Fig. 6.28 Modification of a wind plant control using voltage reference.

In Fig. 6.25 the slow change of the voltage reference is accomplished by the plant controller, in Fig. 6.28 this is done by the washout filter. Under the assumption that the filtered voltage v_{filt} does not change significantly during voltage dips (large time constant T_V) both implementations are identical with respect to the reactive current injection. Only the slow reference value Δv_{WT_i} will be slightly different as it has to compensate the difference between v_{filt} and v_{WT0_i}. But this difference is without practical implications.

Var Control scheme by setting the reactive power reference of individual wind turbines using the plant Var controller

The implementation of the current reference control at turbine level can be accomplished as shown in Fig. 6.29. The reactive current reference provided by the WPP controller is Δq_{WT_i} This value is added to the value q_{WT0_i} which is considered as the steady state setting at the wind turbine. The reactive current is calculated by dividing the reactive power reference by a filtered voltage value. A filtering of the voltage is necessary to prevent that the steady state control counteracts the fast voltage control. Assuming an underexcited operation before a grid fault, without a filter the reactive current would increase in linear fashion with a decrease of voltage.

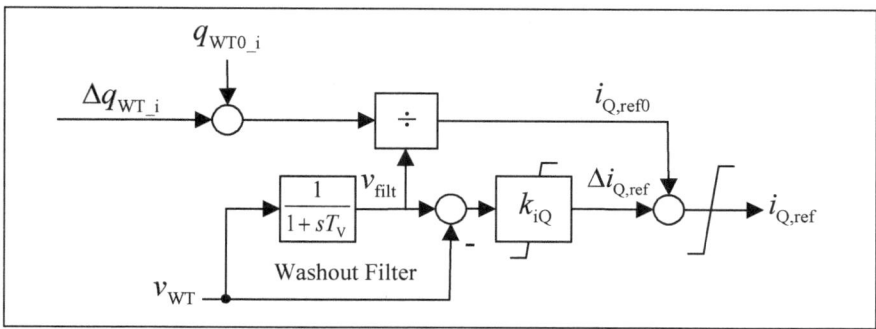

Fig. 6.29 Wind turbine Var control using reactive power reference.

Var Control scheme by setting the power factor reference of individual wind turbines using the plant Var controller

A power factor control can be implemented in a comparable fashion the reactive power control. A power factor reference $\Delta\varphi_{WT\_i}$ is provided by the WPP controller. This value is added to the value φ_{WT0_i} which is considered as the steady state setting at the wind turbine. The reactive power reference is calculated as described in (6.30)

$$q_{\mathrm{WT\_i}} = p_{\mathrm{WT\_i}}^* \cdot \tan\left(\varphi_{\mathrm{WT0\_i}} + \Delta\varphi_{\mathrm{WT\_i}}\right) \qquad (6.30)$$

As shown in Fig. 6.30, both the local voltage at the turbine v_{filt} and the power at the wind turbine $p_{\mathrm{WT\_i}}$ need to be filtered so the calculation of the reactive current reference from the WPP does not interfere with the fast voltage control during grid faults.

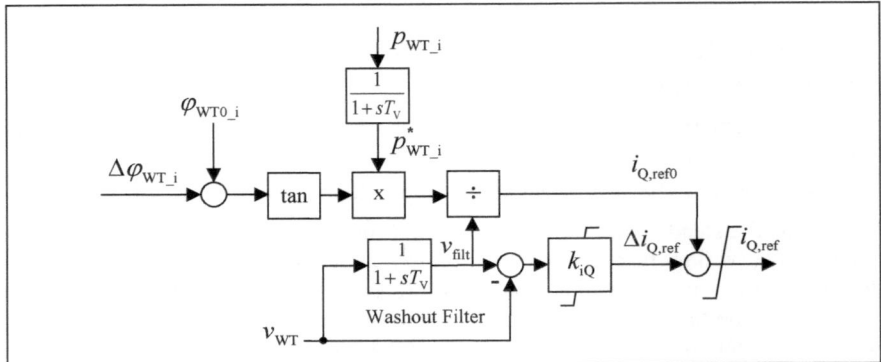

Fig. 6.30 Wind turbine Var control using power factor reference.

Power Factor reference is not used much anymore since this control mode does not allow reactive power control at very low or zero active power.

Evaluation of Differences of the Proposed Control Structures

There are two key differences between a wind plant structure with voltage reference and a wind plant structure with current (or reactive power or power factor) reference:

(1) the output of the plant controller has to be scaled to a different value, depending on the requested reference value

(2) there is an additional filter for the turbine voltage that is not needed using voltage reference.

The scaling of the reference value in the plant controller does not have an impact on the control response. But the filter at turbine level does have some influence in wind plants.

If all turbines in a wind plant had the same voltage at the terminals, there would not be any relevant difference. But due to resistances and impedances of the wind plant collector system, wind turbines close to the PCC will have a different voltage compared to turbines at the end of a cable.

In the case of a reactive power reference, under steady state conditions all turbines will supply the same reactive power – independently from the voltage.

At capacitive (overexcited) operation as an example, a turbines closer to the PCC will operate at a lower voltage than the last turbine at the end of the line.

In the case of voltage reference, if there is a change of voltage between different turbines, due to the control law $i_{QWT,ref} = k_{iQ} \left(v_{WT,ref} - v_{WT} \right)$, turbines at the end of a line with a higher voltage would automatically provide less reactive power because the voltage difference $v_{WT,ref} - v_{WT}$ is smaller. As a result, turbines close to the PCC will provide more reactive power than those at the end of the line. This reduces the risk of over- or undervoltage for turbines at the end of a line and also leads to a slight reduction of cable losses in the wind plant.

6.5 Evaluation of the Proposed Reactive Power Control Structure

Wind plants connected to the high voltage grid are substituting conventional power stations using synchronous generators. For the analysis of the impact of replacing a conventional 50 MW power station, the following systems are compared

(1) a wind plant using the proposed control structure with voltage reference at turbine level.

(2) a synchronous generator plant

(3) a wind plant using reactive power control and a dead band according to [51]

In a *first step*, a wind plant with the proposed control structure using a voltage reference is compared to a synchronous generator. In a *second step*, a short overview over of the grid code development that lead to the existence of a deadband in existing wind turbine reactive power control structures is given In a *third step*, a wind plant using the proposed control structure is compared to a wind plant using existing control structures with a dead band during grid faults.

The producer oriented sign convention is applied for all simulation figures, generator active power and overexcited reactive power are considered to be positive.

6.5.1 Comparison of Wind Plant and Synchronous Generation

The reaction of a wind plant using voltage reference at turbine level with a DFG system according to Fig. 6.31 is compared to a conventional generator system with the same rating (see Fig. 6.10). Both systems are compared at the 110 kV high voltage terminals of the transformer.

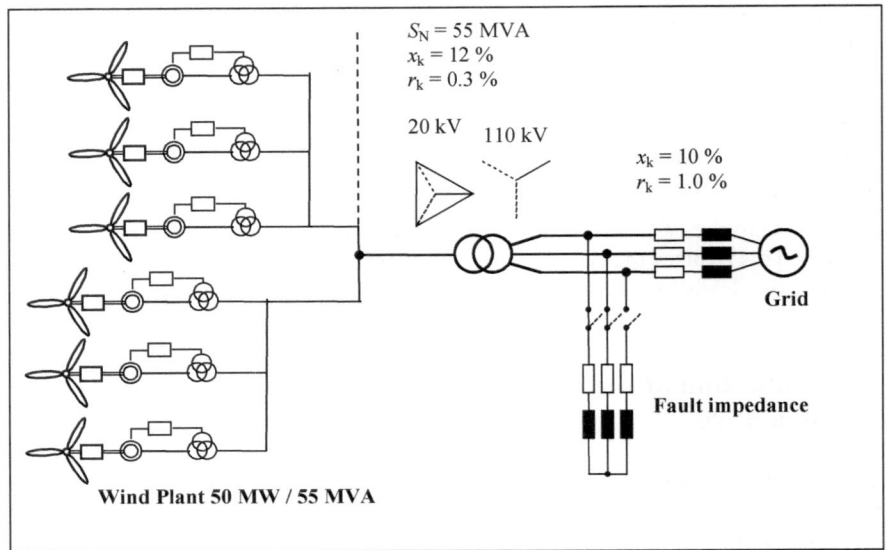

Fig. 6.31 Configuration of 50 MW wind plant.

The wind plant controller is operated using a voltage static with a reactive current gain $k_{iQWP} = 2$ p.u. at the HV terminals. The wind turbines are operated with a reactive current gain $k_{iQ} = 2$ p.u. with a control point at the HV terminals using the control law described by (6.29). The resulting effective reactive current gain at the LV terminals is $k_{iQWT,ref} \approx 4$ p.u..

The wind turbine models used are based on the generic model approaches for the aerodynamic system (section 2), the mechanical structure (section 3), the active power control (section 4) the generator and converter (section 5) and the reactive power control described in this section.

The simulations of the synchronous generator is based on the model as described in section 6.3.4.

The following cases are simulated:

■ close and distant grid fault with effective voltage drops to 15% and 80% at the high voltage terminals. This value would apply if no reactive current would be fed in by the wind plant or the generator during the fault. The reactive current in-feed will effectively lead to higher voltages at the terminals during the fault.

■ a phase angle jump by 10 deg at the high voltage terminals

■ voltage step changes by +5% and -5% in the high voltage grid

Response to Grid Fault close to the High Voltage Terminals

The reaction to a 150 ms grid fault close to the high voltage terminals for a wind plant and a synchronous generator is shown in Fig. 6.32.

The active power of the wind plant returns to rated power within around one second but is almost constant afterwards. The active power of the synchronous generator returns much faster to rated power, but oscillates heavily for several seconds following the fault. The details of the active power recovery of the wind plant may vary depending on the turbine control, but larger oscillations can generally be avoided.

The reactive power at the beginning of the grid fault is considerably smaller for the wind plant. Even though the reactive power of the synchronous generator is more than twice as high, there is close to no increase in grid voltage.

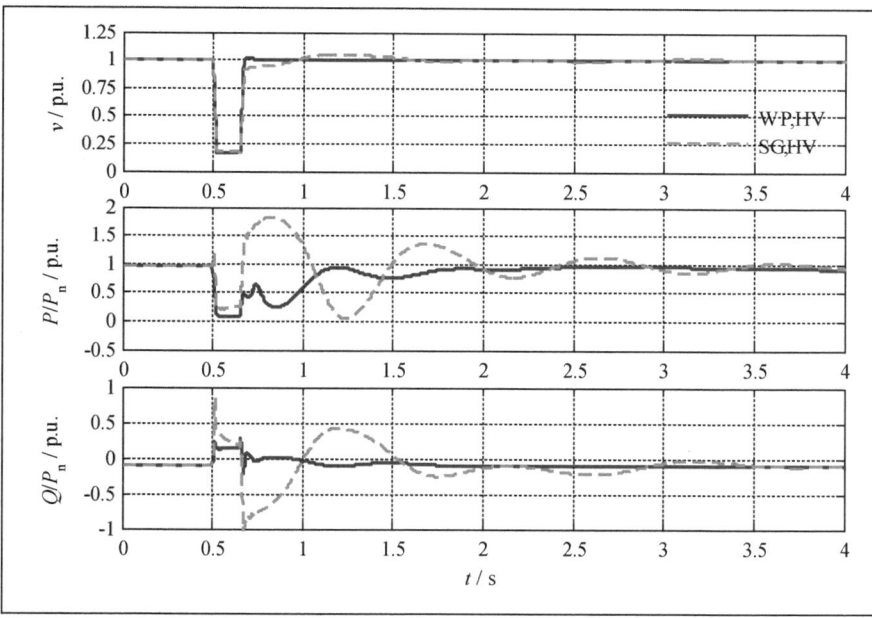

Fig. 6.32 Voltage drop down to 20% rated voltage, comparison of 50 MW wind plant (solid line) and synchronous generator (dashed line).

A closer look at the fault event is given in Fig. 6.33 for the HV terminals and in Fig. 6.34 for the MV terminals. The initial reactive current at HV level of the wind plant is less than 50% of the current of the synchronous generator. Still the voltage can only be increased by 2% compared to the wind plant.

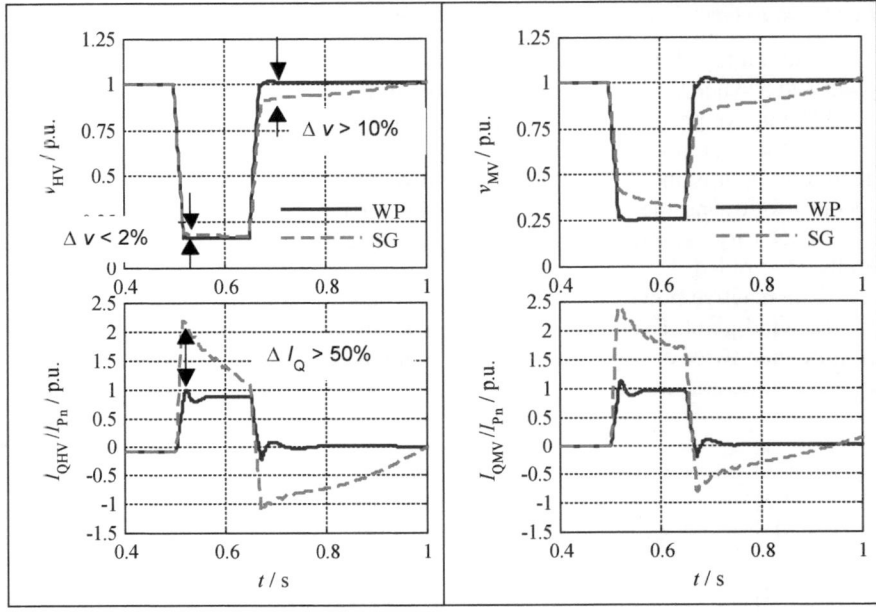

Fig 6.33 HV-terminals: voltage drop down to 20% rated voltage, comparison of 50 MW wind plant and synchronous generator.

Fig. 6.34 MV-terminals: voltage drop down to 20% rated voltage, comparison of 50 MW wind plant and synchronous generator.

Following the fault, reactive power and voltage are close to constant for the wind plant. In the first 300 ms following fault clearing, the synchronous generator it is consuming reactive power, as a result the grid voltage is reduce by almost 10%. The reactive power of the synchronous generator oscillates for several seconds as result of oscillations of the rotor.

Response to Distant Grid Fault

The reaction to a 150 ms grid fault distant to the high voltage terminals for a wind plant and a synchronous generator is shown in Fig. 6.35.

The active power of the wind plant oscillates slightly for about one second following the fault. The active power during the fault may depend on turbine rating and turbine control setting. An active power reduction is possible in case of a smaller rating of the converter, but the power recovery would not lead to relevant oscillations. Even though the voltage drops only by 15% at the high voltage terminals, the active power of the synchronous generator oscillates for several second with an amplitude of up to 0.3 p.u..

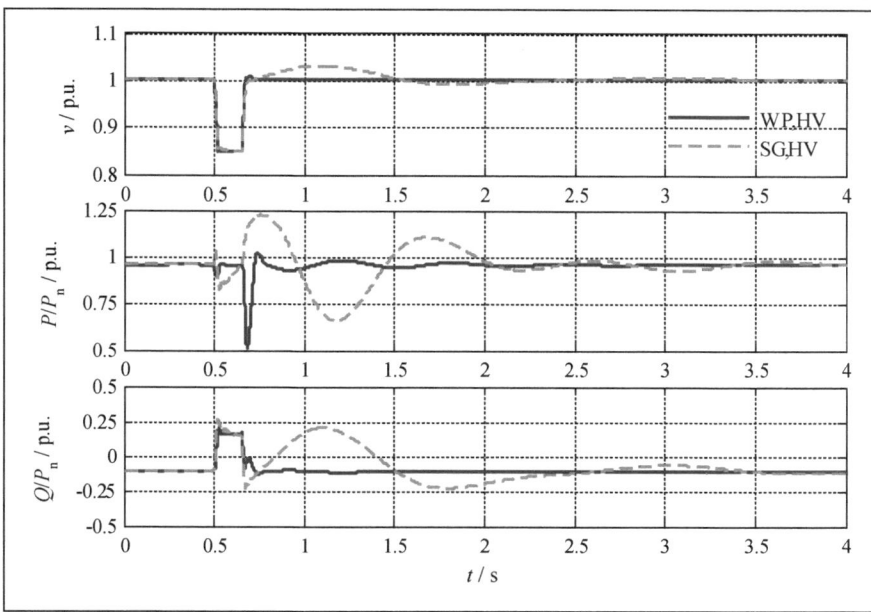

Fig. 6.35 Voltage drop down to 85% rated voltage, comparison of 50 MW wind plant (solid line) and synchronous generator (dashed line) at the high voltage terminals.

During the fault, the reactive power response of the wind plant is equivalent to the synchronous generator. Following the fault, the reactive power of the synchronous generators oscillates for several seconds, the wind plant reactive power reaches its final value within 100 ms after fault clearing.

Response to Voltage Phase Angle Step Changes

The voltage phase angle can change very fast following switching events in the grid that change the reactive load of the system. The response of a wind plant and a synchronous generator to a phase angle change of 10 deg in the high voltage grid is shown in Fig. 6.36.

The wind plant is hardly affected by the phase angle change. Active and reactive power change by less than 10% for less than 50 ms. The phase angle is constant following the step change. By contrast, the active power of the synchronous generator changes by more than 40% and oscillates for several seconds. As a result of the active power oscillation, also the phase angle changes proportional to the active power changes for several seconds.

It shows that although a wind plant could trigger oscillations in the grid, it will not be part of any power swing or phase angle swing in the grid under normal grid conditions.

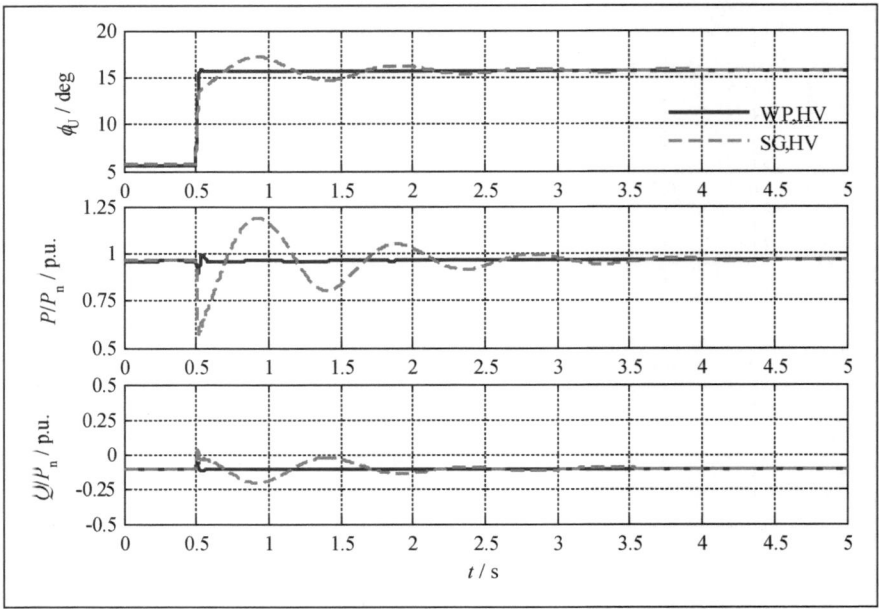

Fig. 6.36 Phase angle jump by 10 deg: comparison of 50 MW wind plant (solid) and
synchronous generator (dashed) at the high voltage terminals.

Response to Voltage Step

The response to a step change in voltage for a wind plant and a synchronous generator is shown in Fig. 6.37 for the reactive current and in Fig. 6.38 for the reactive current gain k_{iQ} directly following the voltage step change; before possible control action of the generator AVR can take place. The values for the gain are shown for the synchronous generator and for the wind plant on HV-level and at turbine level on the MV and LV side.

A step response that is equivalent to a synchronous generator can be achieved for a reactive current gain at HV-level of around 2 p.u. The corresponding equivalent reactive current gain at turbine level at MV-side is 3 p.u., at LV-side 4 p.u. (see Fig. 6.38).

The shown reactive current gain is the minimum gain necessary to achieve a response equivalent to a synchronous generator. A higher gain would lead to an improved response compared to a synchronous generator.

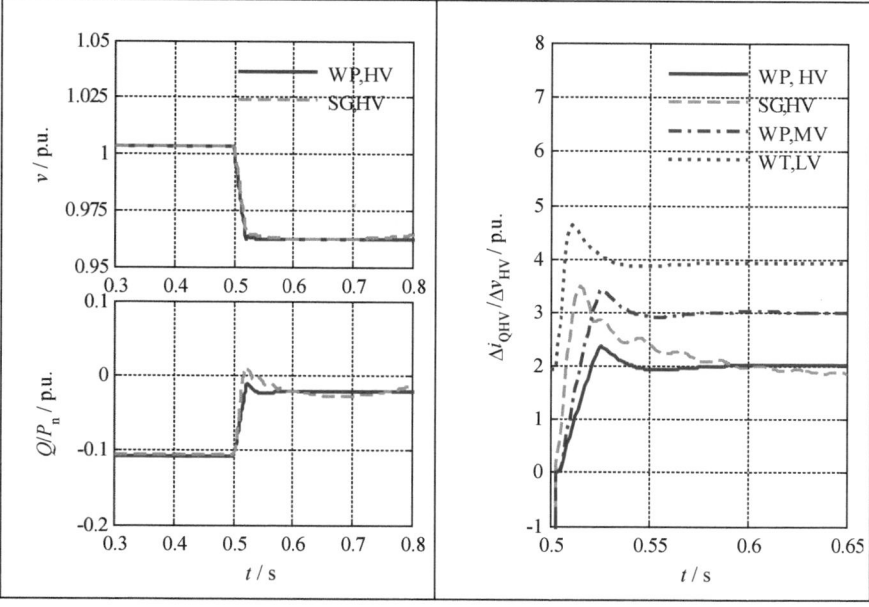

Fig. 6.37 Voltage and reactive power of a wind plant (solid) and a synchronous generator (dashed) at the high voltage terminals following a voltage drop.

Fig. 6.38 Reactive current gain $k_{iQ} = \Delta i_Q / \Delta v$ of a wind plant (solid) and a synchronous generator (dashed) following a voltage drop. The effective reactive current gain is also shown for the wind turbines at MV level (dash-dot) and at LV level (doted).

6.5.2 Background on Reactive Power Control of Wind Turbines with Dead Band during Grid Faults

Development of Voltage Control Requirements for Wind Turbines during Grid Faults

The first grid operator to require voltage control during grid faults for wind turbines was E.ON Netz in 2003 [62]. The decision was based on previous discussions between grid operators and turbine manufacturers in 2002. In a study of E.ON Netz presented in 2003 on the impact of wind turbines on the grid, [64], a desired impact of supporting the grid voltage by using voltage control at turbine level (Fig. 6.39) is shown.

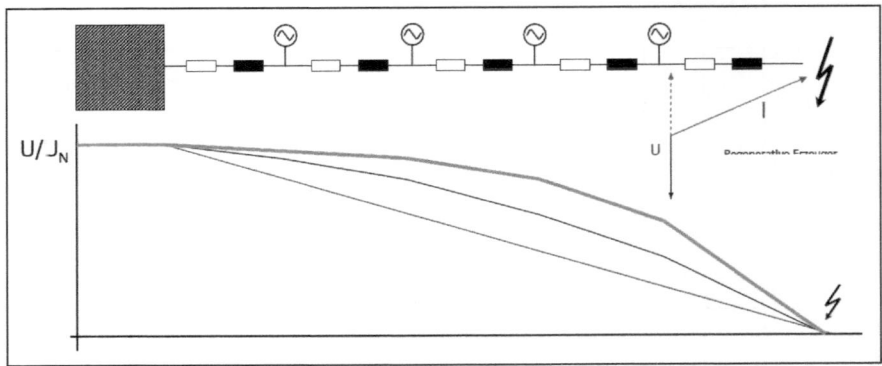

Fig. 6.39 Desired effect of voltage control at turbine level is to support the voltage during grid faults, thin line: no voltage support, thick line: with voltage support. Source: E.ON Netz [64].

A comparable analysis of Vattenfall Europe Transmission came to similar results, it was expected that voltage control of wind turbines would reduce the area affected by a grid fault ([74], see Fig. 6.40 and Fig. 6.41).

Based on a simple grid model the following relation between voltage phase angle change and voltage magnitude was estimated by E.ON during grid faults for synchronous generators [64]. As a result, a gain of the voltage control of $k_{iQ} = \Delta i_Q / \Delta v$ of 2 p.u. was demanded.

What had not been sufficiently analyzed at that time was the location at which this requirement should be valid. The grid operator demand was the PCC, the wind turbine manufacturers position was the converter terminals. This was due to the fact that a sufficiently fast control response was only possible if no communication delay between the measurement point (at the PCC) and the location of the converter (commonly at the LV terminals) needed to be considered.

Technically, the position of the manufacturers is correct. But moving the point of control has an impact on the effective gain at the HV terminals (see Fig. 6.38). A reactive current gain of 2 p.u. at the HV terminals would require a gain higher than 2 p.u. at the MV terminals and even higher gain at the LV terminals of the wind turbine to have the same impact at the HV terminals. This until now has not been reflected in German grid codes.

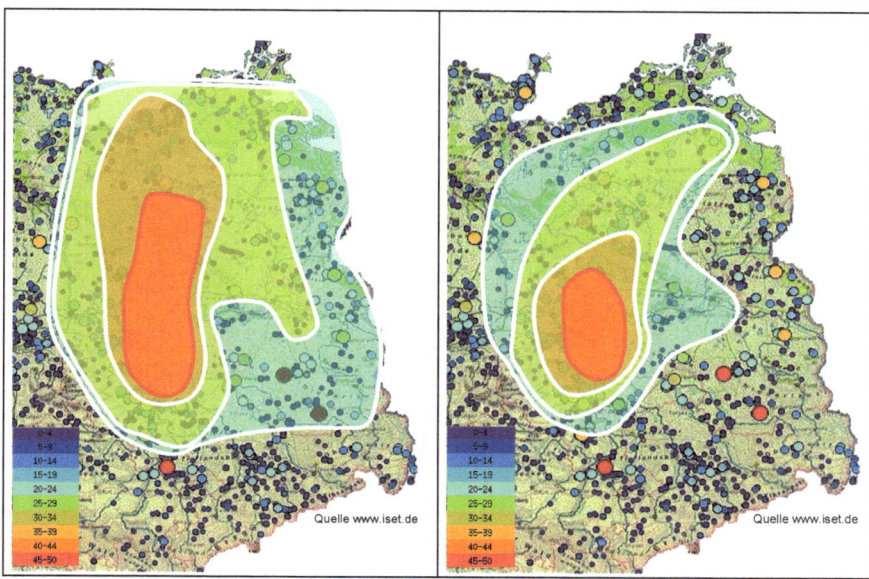

Fig. 6.40 Area with reduced voltage affected by grid fault without voltage support by wind turbines. Source: Vattenfall Europe Transmission [74].

Fig. 6.41 Area affected by grid fault using wind turbines with voltage control during the fault. Source: Vattenfall Europe Transmission [74].

Reactive Power Reference at Turbine Level using a Dead Band during Voltage Dips

Wind turbines in 2001 .. 2003 were usually connected to the distribution system, the size of the turbines (sometimes only several kW) and its commonly distributed location did not justify a connection to the transmission system. As a consequence, the distribution grids that used to have very little production dealt with wind turbines as 'negative' loads, with power factor control as required method to control reactive power. Most turbines were required to operate with a fixed power factor [23].

In order for the new requirements to be easily compatible with power factor control, voltage control was only required outside the standard operating range in the distribution system.

This resulted in the definition of a 'dead band' around rated voltage, where power factor control was active. Outside the standard operating range, voltage control was required. The resulting control law for the reactive current is described in(6.31), see also Fig. 6.42.

$$i_{Q,ref} = \begin{cases} \dfrac{p}{v}\sin(\varphi_{ref}) & \text{for } 0.9 \le v \le 1.1 \\ -2\cdot(v-\overline{v}_0) & \text{for } v < 0.9 \text{ and } v > 1.1 \end{cases} \qquad (6.31)$$

with $\overline{v}_0$ as the average voltage before the fault.

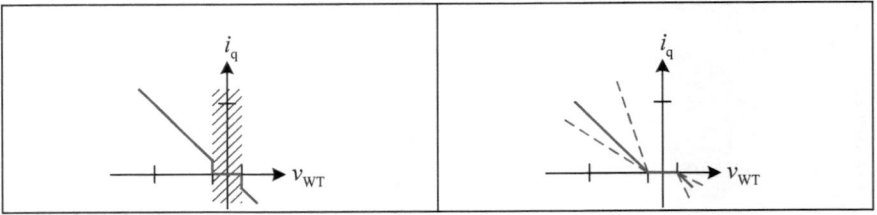

Fig. 6.42 First version of reactive current demand during grid faults by E.ON Netz [23]. **Fig. 6.43** Modified reactive current demand during grid faults [51], [68].

The specification according to (6.31) leads to discontinuities at 0.9 p.u. voltage and to a possibly counter-productive result if the turbine had been operating in overexcited mode before a fault happened. In this case, the reactive current contribution could actually be reduced during the fault.

A modified requirement according (6.32) is described in [51] and [68] that avoids discontinuities (see Fig. 6.43):

$$i_{Q,ref} = \begin{cases} \dfrac{p}{v}\sin(\varphi_{ref}) \text{ or } \dfrac{q_{ref}}{v} & \text{for } 0.9 \le v \le 1.1 \\ k_{iQ}\cdot(v-\overline{v}+\Delta v_0)+\overline{i}_{Q0} & \text{for } v < 0.9 \\ k_{iQ}\cdot(v-\overline{v}-\Delta v_0)+\overline{i}_{Q0} & \text{for } v > 1.1 \end{cases} \qquad (6.32)$$

with $k_{iQ} = \Delta i_Q / \Delta v$ as gain, $\overline{i}_{Q0}$ and $\overline{v}_0$ as average reactive current and average voltage before the fault respectively and Δv_0 as dead band of 0.1 p.u. in this case. Different static values of the gain k_{iQ} and the dead band may be selected.

6.5.3 Comparison to Wind Plant using Reactive Power Control with Dead Band

Response to Grid Fault close to the High Voltage Terminals

The reactive current capability of wind turbines is commonly limited to 1 p.u. for the duration of a grid fault as demanded for example in [51] and [68]. Therefore the response of wind turbines with different control strategies to an electrically close grid fault does not show significant differences during the fault. The maximum reactive current of 1 p.u. at the turbine terminals is fed in during the fault. For stiff grids the response is equivalent to the one shown in Fig. 6.32 - Fig. 6.34.

Under very weak grid conditions, the response during and following voltage recovery of a continuous control (without dead band) as proposed will have advantages. The analysis of weak grids is beyond the scope of this work.

Response to Distant Grid Faults

In Fig. 6.44 the reactive currents during a distant fault are compared for a synchronous generator (dashed), a wind plant with continuous voltage control (solid) and a wind plant according to (6.32) with a dead band.

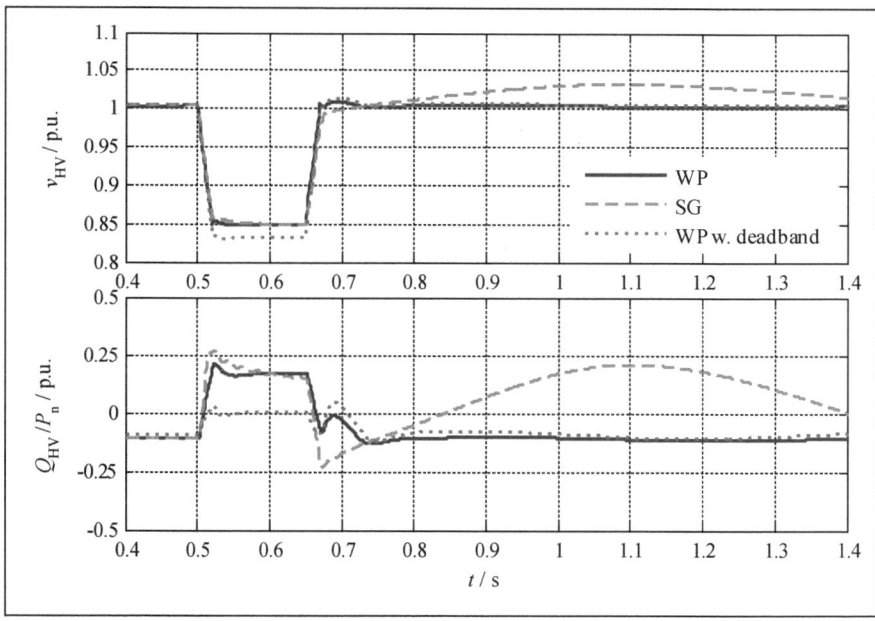

Fig. 6.44 Reactive currents at the HV-terminals for a wind plant with voltage reference (solid), for a synchronous generator (dashed) and for a wind plant with reactive power reference using dead band (doted).

Fig. 6.45 shows the effective voltage control gain during the fault at the PCC at HV-level. A control with dead band shows only a very small effective gain compared to the other systems.

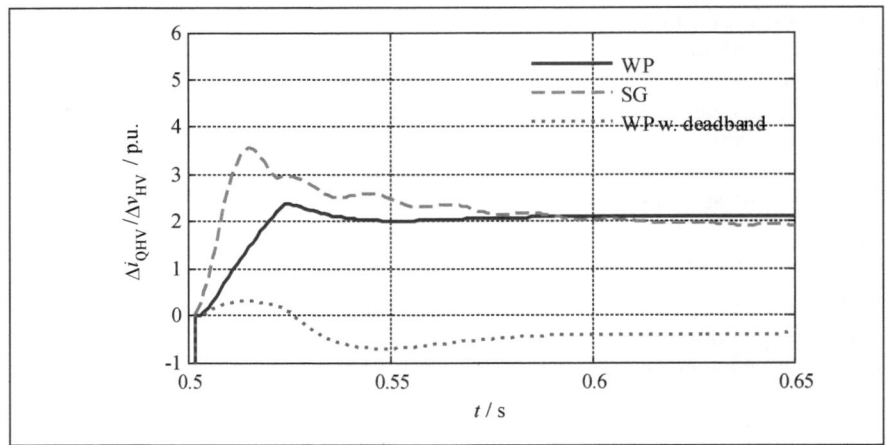

Fig. 6.45 Effective reactive current gain $k_{iQHV} = \Delta i_{QHV}/\Delta v_{HV}$ for a distant grid fault of a wind plant using voltage reference (solid), a synchronous generator (dashed) and a wind plant using reactive power reference with dead band (doted). Voltage and reactive power are shown for the HV side of the HV-transformer.

Response to Voltage Steps

The response to a step change in voltage at the high voltage terminals is shown in Fig 6.46 for the reactive power and in Fig. 6.47 for the corresponding effective reactive current gain $k_{iQHV} = \Delta i_{QHV}/\Delta v_{HV}$. In the case of voltage reference, the wind turbines immediately respond to the voltage change. The plant controller is only need for a fine-tuning of the reactive power reference. As a result, the reactive power response is very fast.

In the case of a reactive power reference, the turbines do not directly respond to the voltage changes due to communication delays between the plant controller and the turbines. In contrary, because the turbines had been operating with an inductive reactive power reference before the voltage drop, the reactive current actually increases in the first 100 ms following the voltage decrease (see doted curve in Fig. 6.46). The turbine controller needs to increase the reactive current to maintain constant reactive power. This leads to a negative reactive current gain during the first 100 ms following the grid fault (see Fig. 6.47) – actually destabilizing the grid voltage even further during this period.

Once the modified reference signals of the plant controller have reached the turbine, the reactive power changes. The rate of change is limited due to communication delays and in order to avoid instability of the control. The time constants of the plant controller must be considerably slower than those of the turbine controller.

The response to the voltage change in the case of local voltage control is not only faster, it can also effectively limit the voltage change at the PCC. In the case

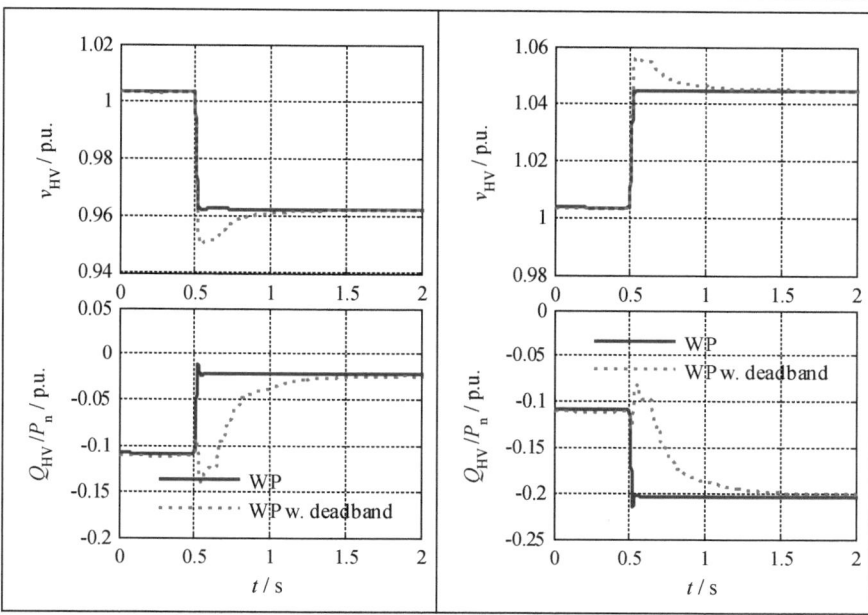

Fig. 6.46 Reaction to voltage step changes of a wind plant using voltage reference (solid) and a wind plant using reactive power reference with dead band (doted). Voltage and reactive power are shown for the HV side of the HV-transformer.

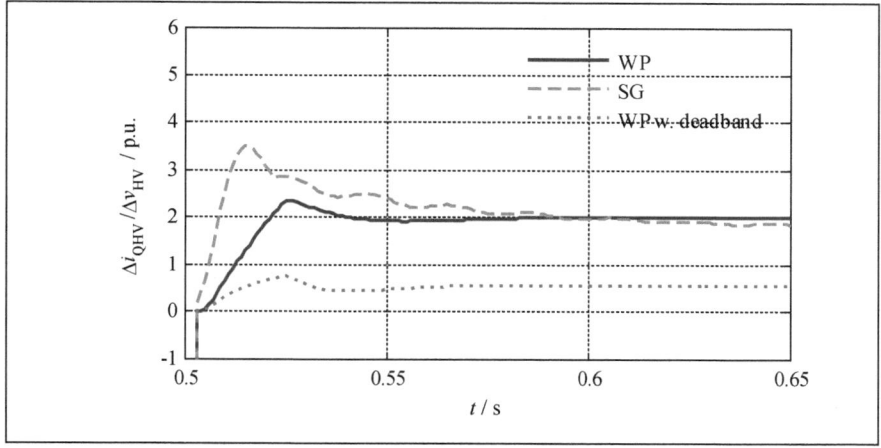

Fig. 6.47 Effective reactive current gain $k_{iQHV} = \Delta i_{QHV}/\Delta v_{HV}$ for a voltage step change of 5% of a wind plant using voltage reference (solid), a synchronous generator (dashed) and a wind plant using reactive power reference with dead band (doted).

of a reactive power reference, an initial increase in voltage following the voltage step cannot be avoided due to communication delays within the wind plant.

6.5 4   Comparison to Measurements

A fast change of voltage within the normal operating range as is could be expected from a switching operation in the grid has been reproduced by the measurement setup in Fig. 6.48. Initially, the turbine operates at an underexcited operating point. At t = 1 s the switch S1 is opened. This leads to a step change of the voltage.

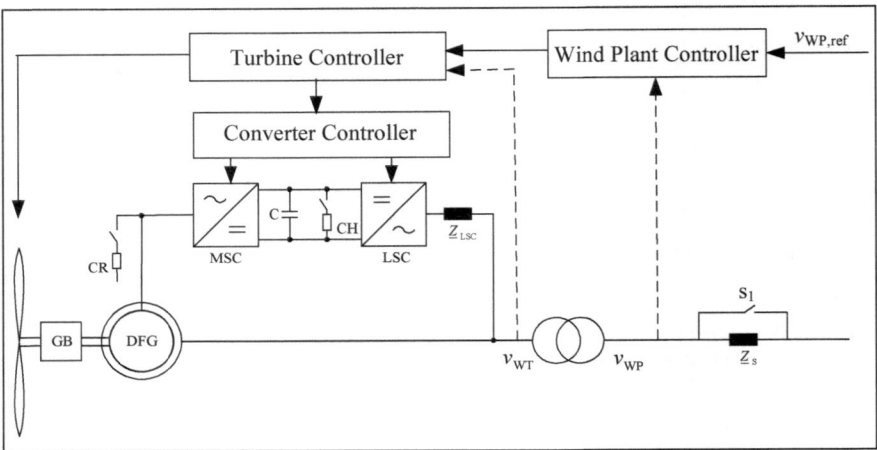

Fig. 6.48 Measurement setup used to measure voltage step changes.

Fig. 6.49 shows the measurement of voltage, active power and reactive power. following a switching event at t = 1 s and a second event 10 s later that closes switch 1 again. The fast response of the voltage control limits the voltage change to about 4 %. The reactive control gain of the voltage static at MV level calculated from as

$$k_{iQWP} = \frac{i_{QWP,ref}}{v_{WP,ref} - v_{WP}} \tag{6.33}$$

remains constant throughout the measurement.

The response of a wind plant of 10 turbines to a voltage step change is shown in Fig. 6.50. The voltage is changed in several steps by forcing the OLTC tap settings. The wind plant is operated with Q(V) control (voltage static) at the PCC. Voltage droop at plant level is not activated, therefore the reactive power at the PCC does not immediately reach the final reference value following a voltage step.

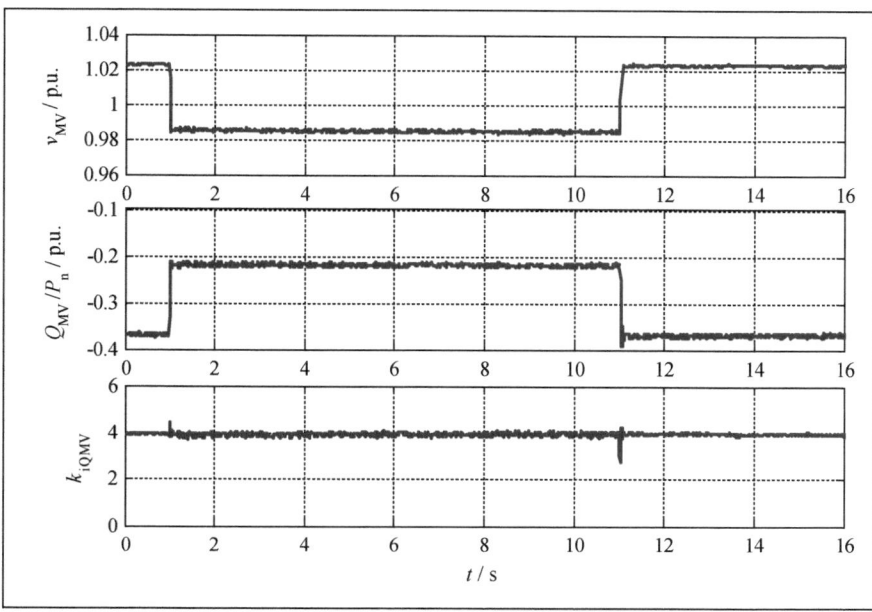

Fig. 6.49 Measured step change in voltage and reactive power for a 2 MW wind turbine.

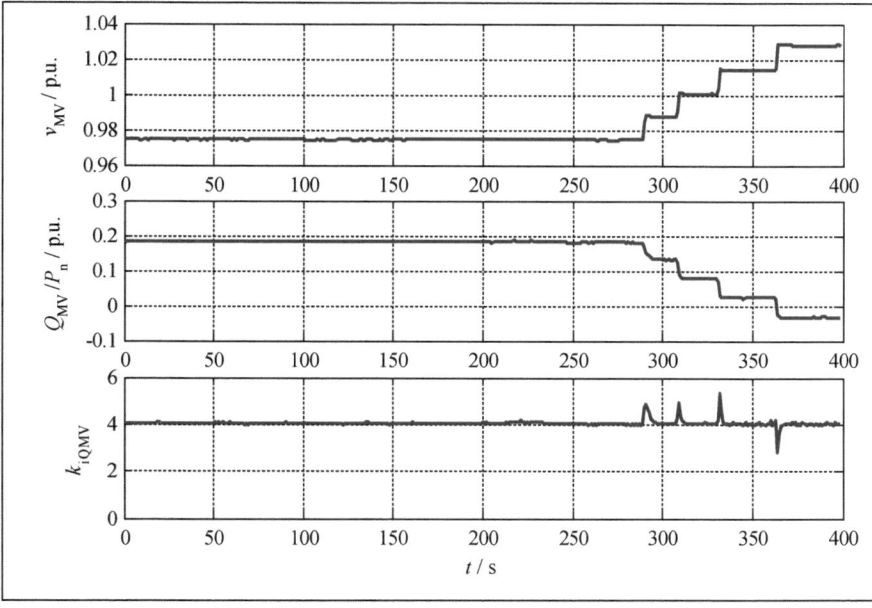

Fig. 6.50 Measured step change in voltage and reactive power for a wind plant.

6.6 Summary

In a first step, the requirements for reactive power control in the grid have been analyzed. It can be shown that the requirements on control speed depend on the situation. In the case of slow changes of the voltage and in case of grid operator reference changes, a slow response of the control is preferable. The normal reason for slow changes are changes in the grid load flow. In the case of fast voltage changes, a fast reactive current response is required.

In a second step, the response of a synchronous generator to grid faults and voltage step changes is analyzed. It can be shown that over the duration of a grid fault the effective reactive currents are mainly defined by the transient reactance x_d'. As a result, a close to linear correlation between voltage decrease and reactive current increase can be deduced. This proportional reactive current gain $k_{iQHV} = \Delta i_{QHV}/\Delta v_{HV}$ at the high voltage terminals is at least 2 p.u. for typical synchronous generators.

Based on the analysis a control structure for the reactive current control of wind turbines and wind plants is derived that provides the desired control response. At turbine level, a proportional reactive current gain is required as response to fast voltage changes. This reactive current must be in addition to the pre-fault steady state reactive current.

Two control structures for wind plants are derived. They are both based on a slower outer control loop at plant level to control load flow and a fast control loop at turbine level to respond to voltage changes. The first structure is based on voltage reference values by the wind plant controller, the second structure is based on reactive power or current reference values by the plant controller. This allows an easy extension of existing wind plant control concepts using reactive power reference.

The response of a wind plant with a control using voltage reference and a synchronous generator are compared for different grid faults and step changes of voltage. The proposed control structure shows a very good response at the 110 kV terminals. Especially for distant faults, the response of the wind farm to grid faults using this control can offer the same or an improved voltage support compared to synchronous generators. In contrast to synchronous generators, wind turbines to not lead to reactive power oscillations and do not absorb reactive power following the voltage recovery.

Older control concepts of wind turbines are using a dead band for voltage control during grid faults as it had been proposed in several grid codes. A comparison of this older concept to both synchronous generators and the proposed control concept shows a clear decrease in voltage support and as a result larger voltage excursions during grid disturbances if a control concept using a dead band is used.

7 Summary and Conclusion

This thesis is focused on two main fields,

(1) the development of a generic wind turbine and wind plant model for the use in grid simulation studies using the minimum number of parameters necessary to represent all relevant effects and that is based on a physical deduction of all relevant parameters

and

(2) the design of a reactive power control system for wind turbines and wind plants based on an analysis of the existing grid that can provide static and dynamic capabilities to ensure a stable voltage and reactive power control for future grids without remaining synchronous generation.

7.1 Generic Wind Turbine Model Development

The requirements for generic models are simplicity, a limited number of parameters and a design that allows the parameterization for different, manufacturer specific implementations.

Wind turbines, on the other side, are fairly complex structures, combining (1) elements of aerodynamics for the calculation of the power input to the turbine, of (2) structural mechanics for calculating the response of rotor, drive train and tower to load changes like grid faults, (3) the control and design of generator and converter to derive the current limits and the current response to fast voltage changes, and (4) a nonlinear control design for the turbine controller to cope with the nonlinearities given by the blades and the design limitations of turbine and converter.

The general aim for the generic model design is therefore not to maximize the achievable level of accuracy using a highly complex model or a combinations of models. The aim in this thesis is to find the minimum level of detail that is necessary to model each component of a wind turbine. The target is to achieve high steady state accuracy and a good approximation during transient events.

A model simplification can be based on empirically or numerically optimized models. Even though good levels of accuracy can be achieved by this approaches, the estimation of parameters is difficult, and the choice of parameter values is not transparent for the user. Therefore, a key focus of this work has been to find model structures that can analytically be derived from the physical structure of the component. Also, all parameters should be derivable from physical quantities.

The key contributions for the development of generic models in this work are:

Generic Aerodynamic Model

The turbine rotor converts the flow of air into mechanical power of the drive train. Turbine aerodynamics are highly nonlinear, and detailed models based on blade element theory are computationally intensive. Common simplified models based on static c_P-λ-Θ tables still need a large number of parameters. As one possiblity of data reduction, functional approximations for c_P-λ-Θ tables have been analyzed and an improved functional representation is derived.

As an alternative, a linearized aerodynamic model has been proposed. Compared to c_P-λ-Θ tables, the linearized approach gives an insight into the aerodynamic system by using physical parameters based on partial derivatives of power with respect to blade angle change and rotor speed change. The model is compared to simulations using c_P-λ-Θ tables and measurements. It is usually accurate enough to replace aerodynamic models based on c_P-λ-Θ tables for grid simulation studies.

Generic Mechanical Model

The analysis of the structural mechanics of the wind turbine shows a large number of degrees of freedom. With respect to grid simulations, the key question is the impact of the mechanical system on the torsional degrees of freedom of the drive train which result in the variation of the generator torque. It can be shown that a reduction of the mechanical structure to a two mass spring-damper system provides a sufficient level of detail for variable speed wind turbines. Higher order eigenfrequencies are not relevant for grid simulation studies of such turbines. The calculation of equivalent drive train parameters based on given data is described.

The drive train eigenfrequency can change depending on wind speed, operating point and mechanical influences like production tolerances and aging effects. This eigenfrequency is usually very well damped. In case it is to be considered in grid studies, instead of using more detailed models the variation of the eigenfrequency in the model is recommended.

Turbine Control Model

The key concepts and limitations of wind turbine active power control are analyzed. Model representations for the pitch – speed control loop and for the torque – speed control loop and for the decoupling of these two loops are derived.

A generic control model is presented that can reproduce different turbine and manufacturer specific parameterizations. An active drive train damper implementation in the model is compared to an increase in damping of the drive train of the model. A difference in power output is visible only for a short period following a grid fault.

Generic DFG and FSC Generator and Converter Model

Variable speed wind turbines are commonly equipped with generators and converters using either a doubly fed generator systems (DFG) or a full size converter (FSC). Based on an analysis of physical equations of the generator and the control design of the converter, a new simplified generator and converter model has been derived for DFG and FSC. The model derived for the DFG is far simpler that existing models based on separate generator and converter models. The key parameters are based on the generator data sheet and basic converter rating information. A very good correlation of simulations with measurements can be shown for both DFG and FSC.

An additionally simplified model of the DFG is analyzed; it provides good accuracy in steady state conditions but is less accurate following fast voltage changes. The generator/converter model does not cover negative sequence control where a large number of manufacturer specific implementations exist that would require implementing many different solutions. The reactive power control model is derived as part of the wind plant reactive control.

The thesis provides generic models for all components of a wind turbine. New modeling approaches are presented for the aerodynamic model and the generator/converter model. The necessary level of detail of the mechanical model and the turbine control model are deduced and reduced systematically to the minimum level required to represent different turbine types and manufacturer specific implementations.

The model presented is more accurate and allows a far better deduction of parameters from their physical equivalent than existing generic models. The work provides a good basis for an increased acceptance and use of generic models for grid simulation at national and international level.

7.2 Wind Plant Reactive Power Control

The development of requirements for renewable sources in grid codes is in a transition phase. Renewable sources like wind power have been recognized as relevant source of power that can have a relevant impact on system stability. As consequence, FRT requirements have been stated that shall limit the impact of grid faults. But grid codes so far implicitly assume a certain base capacity of operating synchronous generation units.

A reactive power control structure for wind turbines and wind plants has been developed and tested that provides the necessary capability to ensure a stable system voltage in the medium and high voltage grid even if no synchronous generation is present any more.

Analysis of Grid Requirements

The grid requirements for a stable steady state and dynamic voltage and reactive power control are deduced. An analysis of the development of grid codes shows clear deficits in the existing grid code requirements. The requirements stated are not sufficient to enable a stable control of the system voltage without a base capacity of synchronous generators. The dynamic reactive power control requirements are irrespective of the point of control. The same requirements apply for example to turbines with LV and with MV control event thought this leads to a different response at the HV terminals. Dynamic reactive current requirements have either not been specified in detail or have been poorly implemented with large deadbands that could actually destabilize the system voltage in normal operation instead of providing additional support.

An analytical approach has been developed that quantifies the capability of synchronous generators to support the grid voltage in the case of a grid fault or a sudden voltage change. The dynamic reactive current gain describes the capability of a generator system to provide reactive current in the case of voltage changes at the high voltage terminals.

Both synchronous generators and wind turbines are connected to the medium voltage system. But due to the wind plant collector system, possibly different high voltage transformers and the distributed nature of wind turbines, the comparison of the response of synchronous generators and wind plants should be at the high voltage level.

It can be shown that the minimum dynamic reactive current gain of a common power station with synchronous generators connected to the high voltage system is around 2 p.u. at the high voltage terminals.

Wind Plant Reactive Power Control

Based on the analysis, a wind plant reactive current control system is designed with the aim of providing an improved dynamic response at the high voltage terminals compared to power stations based on synchronous generators.

The control structure is based on a fast acting voltage controller at turbine level and a slower plant controller that sends out voltage reference signals to the individual turbines. In a further step, an alternative control based on a modification of existing turbine controllers using reactive power reference is deduced that can provide a comparable response.

The proposed control structure is compared to synchronous generators. It can provide the same response in the form of a dynamic reactive current gain as synchronous generators during grid faults and voltage step changes. This is chosen as reference case. By changing the parameterization, an improved voltage stabilization is possible. The control response is far superior to synchronous generators following grid faults and following voltage phase angle changes.

The proposed control structure provides a considerably better dynamic response in the case of grid faults than wind plants with existing reactive power control using a dead band.

The developed control structure has been tested in a large number of wind turbines, it already now provides the reactive power control capabilities needed for future grids with reduced or no remaining synchronous generation.

8 References

8.1 Literature

[1] European Commission: EU climate and energy package, http://ec.europa.eu /climateaction/docs/climate-energy_summary_en.pdf.

[2] EWEA: Pure Power - Wind Energy Scenarios up to 2030. www.ewea.org.

[3] EcoGrid.dk: Ecogrid Phase I Summary report. Steps toward a Danish Power System with 50% Wind Energy. www.ecogrid.dk.

[4] "Generic Type-3 Wind Turbine-Generator Model for Grid Studies", WECC Wind Generator Modeling Group, Version 1.1, Sept. 14, 2006.

[5] M. Pöller and S. Achilles, "Aggregated Wind Park Models for Analyzing Power System Dynamics", DIgSILENT GmbH, www.digsilent.de.

[6] R. Gagnon, G. Turmel, C. Larose, J. Brochu, G. Sybille and M. Fecteau, "Large-Scale Real-Time Simulation of Wind Power Plants into Hydro-Québec Power System", *9th International Workshop on Large-Scale Integration of Wind Power into Power Systems as well as on Transmission Networks for Offshore Wind Farms*, Quebec City, 2010.

[7] F. Vandenberghe, "UCTE Investigation Committee on 28/9/03 Blackout in Italy: Lessons and Conclusions from the 28 September 2003 Blackout in Italy", *IEA Workshop* – 29 March 2004.

[8] UCTE: Final Report: System Disturbance on 4 November 2006, www.entsoe.eu.

[9] S. Øye , "Dynamic stall - simulated as time lag of separation", *Proceedings of the 4th IEA Symposium on the Aerodynamics of Wind Turbines*, McAnulty, K. F (Ed.), Rome, Italy, 1991.

[10] M. Hansen, A. D. Hansen, T. Larsen, S. Øye, P. E. Sørensen, and P. Fuglsang, "Control design for a pitch-regulated, variable speed wind turbine", Risø National Laboratory, Roskilde, Denmark, January 2005.

[11] A. Betz, *Wind Energie und ihre Ausnutzung durch Windmühlen*, Bandenhock and Ruprecht, Göttingen, Germany, 1926.

[12] R. Gasch and J. Twele, *Windkraftanlagen, Grundlagen, Entwurf, Planung und Betrieb*, B. Teubner, 2005.

[13] G. Schmitz, „Theorie und Entwurf von Windrädern optimaler Leistung",
Wissenschaftliche Zeitschrift der Universität Rostock, Jahrgang 1955/56.,
Rostock, Germany, 1956.

[14] T. Burton, D. Sharpe, N. Jenkins and E. Bossanyi, *Wind Energy Handbook*,
John Wiley and Sons, LTD, 2001.

[15] E. Hau, *Windkraftanlagen, Grundlagen Technik, Einsatz, Wirtschaftlich-
keit*, 3. Auflage, Springer 2003.

[16] P. M. Anderson and A. Bose, "Stability Simulation Of Wind Turbine Systems"; *IEEE/PES 1983 Summer Meeting*, Los Angeles, California, July 17-
22 1983.

[17] D. Arsudis, "Doppeltgespeister Drehstromgenerator mit Spannungszwi-
schenkreis-Umrichter im Rotorkreis für Windkraftanlagen"; Ph.D. Dissertation TU Braunschweig, Germany, 1989.

[18] S. Heier, *Grid Integration of Wind Energy Conversion Systems*, 2nd Edition, Wiley, April 2006.

[19] V. S. Pappala, "Application of PSO for optimization of power systems under uncertainty", Ph.D. dissertation, University Duisburg, Germany, 2009.

[20] W. Price and J. Sanchez-Gasca, "Dynamic Performance Analysis of the GE
1.5 MW, 60 Hz Wind Turbine- Generator Model Using a Simplified Aerodynamic Model", GE Energy, 2006.

[21] W. Price and J. Sanchez-Gasca, "Simplified wind turbine generator aerody-
namic models for transient stability studies" *Proc. IEEE PES 2006 Power
Systems Conference and Exposition (PSCE)* Oct. 29-Nov 1. 2006 Atlanta,
GA. 2006.

[22] M. Behnke, A. Ellis Y. Kazachkov, A. Muljadi, W. Price and J. Sanchez-
Gasca, "Development and Validation of WECC Variable Speed Wind Turbine Dynamic Models for Grid Integration Studies", *AWEA 2007 Wind
power Conference*, Los Angeles, June 4-7,2007.

[23] E.ON Netz GmbH, "Netzanschlussregeln- allgemein. Technische und orga-
nisatorische Regeln für den Netzanschluss innerhalb der Regelzone der
E.ON Netz GmbH im Bereich der ehemaligen PreussenElektra Netz GmbH
& Co. KG", Rev: December 1[st,] 2001, Bayreuth, 2001.

[24] National Grid, "The Grid Code", Issue 4, Revision 2, March 22nd 2010,
http://www.nationalgrid.com/uk/Electricity/Codes/gridcode/.

[25] I. Erlich and M. Wilch, "Primary frequency control by wind turbines",
2010 IEEE Power and Energy Society General Meeting, Minnesota, July
25 - 29, 2010.

[26] FGW: Technische Richtlinien für Erzeugungseinheiten Teil 3 Bestimmung der Elektrischen Eigenschaften von Erzeugungseinheiten am Mittel-, Hoch- und Höchstspannungsnetz s. Revision 19, 14. January 2009.

[27] IEC: IEC 61400-27-1 ed. 1, Wind Turbines - Part 27: Electrical simulation models for wind power generation - Section 1, CDV 2013-07-31

[28] J.G. Slootweg, H. Polinder and W.L. Kling, "Reduced order Modelling of Wind Turbines" in T. Ackermann, *Wind Power in Power Systems,* John Wiley & Sons LTD, 2005.

[29] A. D. Hansen, F. Iov, P. E. Sørensen, N.A. Cutululis, C. Jauch and F. Blaabjerg, "Dynamic wind turbine models in power system simulation tool DIgSILENT", Risø National Laboratory, Roskilde, Denmark, August 2007.

[30] H. Snel and J. G. Schepers, "Joint investigation of dynamic inflow effects and implementation of an engineering method", Report ECN-C-94-107, ECN, Petten, Holland, 1995.

[31] C. Lindenburg, "Results of PHATASIII development", *International Energy Agency, 28th Meeting of Experts*, Lyngby, Denmark, 1996.

[32] V. Akhmatov, "Analysis of Dynamic Behaviour of Electric Power Systems with Large amount of Wind Power", PhD dissertation, Electric Power Engineering, Ørsted-DTU, Denmark, 2003.

[33] T. G. van Egeln and E. L. van der Hooft, "Dynamic Inflow Compensation for Pitch Controlled Wind Turbines", *European Wind Energy Conference*, London, Nov. 22-25, 2004.

[34] IEEE Working Group on Prime Mover and Energy Supply Models for System Dynamic Performance Studies, "*Hydraulic Turbine and Turbine Control Models for Dynamic Studies*," IEEE Transactions on Power Systems, Vol.7, No.1, February, 1992, pp. 167-179.

[35] Stig Øye, FLEX5 Simulation Software, http://www.dtu.dk/English /Service/Phonebook.aspx?lg=showcommon&type=person&id=3561.

[36] Bladed Wind Turbine Design Software, GL Garrad Hassan, http://www.gl-garradhassan.com/en/GHBladed.php.

[37] Repower MM 100 product description. www.repower.de.

[38] J. Fortmann, "Validation of DFG model using 1.5 MW turbine for the analysis of its behavior during voltage drops in the 110 kV grid", *4th International Workshop on Large-Scale Integration*, Billund, 2003.

[39] A. D. Hansen, P. E. Sørensen, F. Blaabjerg and J. Bech, "Dynamic modelling of wind farm grid interaction", *Wind Engineering*, 26(4) 2002, pp. 191-208.

[40] K. Thomsen, J.T. Petersen, E. Nim, S. Øye and B. Petersen, "A Method for Determination of Damping for Edgewise Blade Vibrations", *Wind Energy*,Vol.3, Issue 4, pp. 233-246, Wiley & Sons LTD, 2000.

[41] T. Thiringer and J. A. Dalberg, "Periodic pulsations from a three-bladed wind turbine", *IEEE Transactions on Power Systems*, Vol. 16, No. 1, 2001.

[42] A. D. Hansen, "Generators and Power Electronics for Wind Turbines", in T. Ackermann, *Wind Power in Power Systems*, Wiley & Sons, Ltd, Chichester, UK, 2009.

[43] M. Pöller, "Doubly-Fed Induction Machine Models for Stability Assessment of Wind Farms", DIgSILENT GmbH, www.digsilent.de.

[44] K. Clark, N. W. Miller and J. Sanchez-Gasca, *Modeling of GE Wind Turbine-Generators for Grid Studies*, Version 4.5, April 16, 2010. GE Energy, 2010.

[45] U. Beckert and P. Stupin, "Dynamisches Verhalten der doppelt-gespeisten Asynchronmaschine mit Berücksichtigung der Stromverdrängung im Läufer", online: http://www.iuz.tu-freiberg.de/~wwwelt/download/HP2_IfE.pdf, TU-Bergakademie Freiberg, Institut für Elektrotechnik, 2005.

[46] C. Jauch, "Stability and Control of Wind Farms in Power Systems", PhD dissertation, Aalborg University, 2006.

[47] A. D. Hansen, P. E. Sørensen, F. Iov and F. Blaabjerg, "Power control of a wind farm with active stall wind turbines and AC grid connection", *European Wind Energy Conference and Exhibition (EWEC)*, Athens, April 20-23, 2006.

[48] S. Engelhardt, „Direkte Leistungsregelung einer Windenergieanlage mit doppelt gespeister Asynchronmaschine", Ph.D Dissertation, Universität Duisburg-Essen, Germany, 2011.

[49] Ian A. Hiskens, "Dynamics of Type-3 Wind Turbine Generator Models.", IEEE Transactions on Power Systems 27.1 (2012): 465-474.

[50] J.G. Slootweg, H. Polinder, W. Kling, "Dynamic modelling of a wind turbine with doubly fed induction generator", IEEE Power Engineering Society Summer Meeting, 2001

[51] Verordnung zu Systemdienstleistungen durch Windenergieanlagen (Systemdienstleistungsverordnung – SDLWindV), BMU, Germany,27.05.2009.

[52] REE – Requisitos de respuesta frente a huecos de tensión de las instalaciones de producción de régimen especial, PO 12.3, November 2005.

[53] FGW: Technical Guidelines for Power Generating Units. Part 4 Demands on Modeling and Validating Simulation Models of the Electrical Character-

istics of Power generation Units and Systems. Revision 4, September 15th, 2009.

[54] FGW: Technische Richtlinien für Erzeugungseinheiten Teil 8 Zertifizierung der Elektrischen Eigenschaften von Erzeugungseinheiten und -anlagen am Mittel- Hoch- und Höchstspannungsnetz, Revision 1, 15. September 2009.

[55] Procedure for Verification Validation and Certification of the Requirements of the PO 12.3 on the response of wind farm in the event of voltage dips, AEE, online: http://www.aeeolica.es/doc/privado/pvvc_v3_english.pdf.

[56] I. Martinez de Alegria, J. L. Villate, J. Andreau, I. Gabilola and P. Ibanez, "Grid Connection of Doubly Fed Induction Generator Wind Turbines: A survey", *European Wind Energy Conference 2004*, http://www,2004ewec. info , London, UK, 2004.

[57] I. Erlich, J. Kretschmann, J. Fortmann, S. Engelhardt and H. Wrede, "Modeling of Wind Turbines based on Doubly-Fed Induction Generators for Power System Stability Studies", *IEEE Transactions on Power Systems*, Volume 22, Issue 3, Aug. 2007 Page(s):909 – 919.

[58] G. Michalke, "Variable Speed Wind Turbines - Modelling, Control, and Impact on Power Systems", Ph.D. dissertation, TU Darmstadt, Germany, 2008.

[59] IEC: IEC 61400-21 ed. 2, Wind turbine generator systems – Part 21: Measurement and assessment of power quality characteristics of grid connected wind turbines.

[60] FGW: Technische Richtlinien für Erzeugungseinheiten Teil 3 Bestimmung der Elektrischen Eigenschaften von Erzeugungseinheiten am Mittel-, Hoch- und Höchstspannungsnetz s. Revision 19, 14. January 2009.

[61] S. Ledwon, F. Kalverkamp, J. Langstädler, B. Schowe-von-der-Brelie, "Dynamic voltage support of decentralized power Generators", *11th International Workshop on Large-Scale Integration of Wind Power into Power Systems as well as on Transmission Networks for Offshore Wind Farms*, Lisbon, Portugal, 2012.

[62] E.ON Netz GmbH, "Grid Code High and Extra high voltage", Rev: August 1^{st,} 2003, Bayreuth, 2003.

[63] J. Machowski, J. W. Bialek and R. Bumby, *Power System Dynamics. Stability and Control*, Second Edition, John Wiley and Sons, 2009.

[64] E.ON Netz GmbH, "Der Einfluss von Windenergie auf das Verbundnetz", Salzburg, 2003.

[65] Eirgrid, SONI, All Island TSO Facilitation of Renewables Studies, June. 2010. online: http://www.eirgrid.com.

[66] BDEW: „Technische Richtlinie Erzeugungsanlagen am Mittelspannungsnetz", Juni 2008.

[67] VDN: „EEG-Erzeugungsanlagen am Hoch- und Höchstspannungsnetz, Leitfaden für den Anschluss und Parallelbetrieb von Eigenerzeugungsanlagen auf Basis erneuerbarer Energien an das Hoch- und Höchstspannungsnetz in Ergänzung zu den NetzCodes", Verband der Netzbetreiber VDN e.V. beim VDEW, 2004.

[68] E.ON Netz GmbH, "Grid Code High and Extra high voltage", Rev. April 1st 2006, online: http://www.eon-netz.com/, Bayreuth, 2006.

[69] National Grid, "The Grid Code", Issue 4, Revision 2, March 22nd 2010, online: http://www.nationalgrid.com/uk/Electricity/Codes/gridcode/.

[70] P. Kundur, *Power System Stability and Control.* New York: McGraw-Hill, 1994.

[71] IEEE, "*Recommended Practice for Excitation System Models for Power System Stability Studies,*" IEEE Standard 421.5-1992, August, 1992.

[72] Matlab, Simulink, SimPowerSystem Toolbox, www.mathworks.com.

[73] Entwurf - Technische Bedingungen für den Anschluss und Betrieb von Kundenanlagen an das Hochspannungsnetz (TAB Hochspannung), FNN im VDE, November 2012.

[74] U. Bachmann, I. Erlich and J. Fortmann, "Auswirkungen verstärkter Windeinspeisung auf das Übertragungsnetz", HusumWind 2003, Husum, Germany, 2003.

[75] Clarke, E., *Circuit Analysis of AC Power Systems*, Vol. 1, New York, Wiley, USA, 1943.

[76] H. Akagi, S. Ogasawara and H. Kim "The theory of instantaneous power in three-phase four-wire systems: A comprehensive approach". *34th Annual Meeting IEEE-IAS*, 1999, pp.431-439.

[77] IEEE Std 1459-2000. *IEEE Trial-Use Standard Definitions for the Measurement of Electric Power Quantities Under Sinusoidal, Nonsinusoidal, Balanced, or Unbalanced Conditions.* IEEE 2000.

[78] J. Niiranen, "About the Active and Reactive Power Measurements in Unsymmetrical Voltage Dip Ride Through Testing", *Nordic Wind Power Conference,* May 22-23, Espoo, Finland 2006.

[79] C. L. Fortescue, "Method of symmetrical co-ordinates applied to the solution of polyphase networks", *34th Annual Convention of the American Institute of Electrical Engineers*, Atlantic City, N.H. June 28, 1918.

[80] T-N. Lê, Kompensation schnell veränderlicher Blindströme eines Drehstromverbrauchers, etz Archiv Bd. 11 (1989) H.8, S.249-253.

[81] C. Feltes, Advanced Fault Ride-Through Control of DFIG based Wind Turbines including Grid Connection via VSC-HVD. PhD dissertation, University of Duisburg-Essen, 2011.

8.2 Publications

[A] J. Fortmann, "Validation of DFG model using 1.5 MW turbine for the analysis of its behavior during voltage drops in the 110 kV grid", *4th International Workshop on Large-Scale Integration*, Billund, 2003.

[B] J. Zeumer and J. Fortmann, „Mögliche Synergien bei Offshore Windparks zwischen WEA-Herstellern und Netzbetreibern", 4. Fachtagung Windtech, Grevenbroich, Germany, 2003.

[C] I. Erlich, J. Kretschmann, J. Fortmann, S. Engelhardt, and H. Wrede, "Modeling of Wind Turbines based on Doubly-Fed Induction Generators for Power System Stability Studies", *IEEE Transactions on Power Systems*, Volume 22, Issue 3, Aug. 2007 pp. 909 – 919.

[D] C. Feltes, S. Engelhardt, J. Kretschmann, J. Fortmann, F. Koch and I. Erlich, "High Voltage Ride-Through of DFG-based Wind Turbines", *2008 IEEE Power and Energy Society General Meeting*, Pittsburgh, 2008.

[E] I. Erlich, J. Kretschmann, S. Engelhardt, F. Koch and J. Fortmann, "Modeling of Wind Turbines based on Doubly-Fed Induction Generators for Power System Stability Studies", *2008 IEEE Power and Energy Society General Meeting*, Pittsburgh, 2008.

[F] I. Erlich, C. Feltes, S. Engelhardt, S. Bauschke, F. Koch and J. Fortmann, „Windenergieanlagen mit doppelt gespeister Asynchronmaschine mit verbessertem Verhalten in Bezug auf neue Netzanforderungen", *etg-/bdew-Tutorial "Schutz- und Leittechnik"*, Fulda, Germany, Nov. 2008.

[G] J. Fortmann, Wilch, M., F. Koch and I. Erlich, "A novel centralised wind farm controller utilising voltage control capability of wind turbines", *PSCC Power System Computation Conference 2008*, Glasgow, July 2008.

[H] I. Erlich, F. Shewarega S. Engelhardt, J. Kretschmann, J. Fortmann and F. Koch, "Effect of Wind Turbine Output Current during Faults on Grid Voltage and Transients Stability of Wind Parks", *2009 IEEE Power and Energy Society General Meeting*, Calgary, 2009.

[I]   C. Feltes, S. Engelhardt, J. Kretschmann, J. Fortmann, F. Koch and I. Erlich, "Comparison of the Grid Support Capability of DFG-based Wind Farms and Conventional Power Plants with Synchronous Generators", *2009 IEEE Power and Energy Society General Meeting*, Calgary, 2009.

[J]   I. Erlich, J. Fortmann, S. Engelhardt and J. Kretschmann, „Spannungsregelung mit moderner WEA-Technik", *Kasseler Symposium Energie-Systemtechnik 2009*, Kassel, Germany, 2009.

[K]   J. Fortmann, S. Engelhardt, J. Kretschmann, C. Feltes and I. Erlich. "Validation of an RMS DFIG simulation model according to new German model validation standard FGW TR4 at balanced and unbalanced grid faults." *8th International Workshop on Large-Scale Integration of Wind Power into Power Systems as well as on Transmission Networks for Offshore Wind Farms*, Bremen, Germany. 2009.

[L]   S. Engelhardt, C. Feltes, J. Fortmann, J. Kretschmann and I. Erlich, "Reduced Order Model of Wind Turbines based on Doubly-Fed Induction Generators during Voltage Imbalances", *8th International Workshop on Large-Scale Integration of Wind Power into Power Systems as well as on Transmission Networks for Offshore Wind Farms*, Bremen, Germany, 2009.

[M]   R. Bluhm and J. Fortmann, "Advanced two-level voltage control in Wind Farms with Doubly Fed Induction Generators", *European Wind Energy Conference 2010*, Warsaw, Poland, 2010.

[N]   J. Fortmann, S. Engelhardt, J. Kretschmann, M. Janßen, T. Neumann and I. Erlich, "Generic Simulation Model for DFG and Full Size Converter based Wind Turbines", *9th International Workshop on Large-Scale Integration of Wind Power into Power Systems as well as on Transmission Networks for Offshore Wind Farms*, Quebec City, 2010.

[O]   J. Fortmann, "New Generic Aerodynamic Model for RMS Simulation of variable speed Wind Turbines", *9th International Workshop on Large-Scale Integration of Wind Power into Power Systems as well as on Transmission Networks for Offshore Wind Farms*, Quebec City, 2010.

[P]   L. Cai, J. Fortmann, R. Blum and I. Erlich, "Power System Dynamic Voltage Stability Analysis for Integration of Large Scale Windparks", *9th International Workshop on Large-Scale Integration of Wind Power into Power Systems as well as on Transmission Networks for Offshore Wind Farms*, Quebec City, 2010.

[Q] C. Feltes, S. Engelhardt, J. Kretschmann, J. Fortmann and I. Erlich, "Dy-
namic performance evaluation of DFIG-based wind turbines regarding new
German grid code requirements." *IEEE Proc. Power and Energy Society
General Meeting.* 2010.

[R] J. Fortmann, S. Engelhardt, J. Kretschmann, C. Feltes, M. Janßen, T. Neu-
mann and I. Erlich, "New Generic Generator Model for RMS Simulation of
Wind Turbines with DFIG or Full Size Inverters", *German Wind Energy
Conference DEWEK 2010*, Bremen, Germany 2010.

[S] L. Cai, Jens Fortmann, R. Bluhm, I. Erlich, "Improvement of Power Sys-
tem Dynamic Voltage Stability by Means of Controlling the Doubly-Fed
Induction Generators with Ancillary Service Bonus Certificate", *German
Wind Energy Conference DEWEK 2010*, Bremen, Germany 2010.

[T] S. Engelhardt, J. Kretschmann, J. Fortmann, F. Shewarega, I. Erlich and C.
Feltes, "Negative sequence control of DFG based wind turbines." *Power
and Energy Society General Meeting, 2011 IEEE.* IEEE, 2011.

[U] J. Fortmann, L. Cai, S. Engelhardt and J. Kretschmann, "Wind turbine
modeling, LVRT field test and certification." *Power and Energy Society
General Meeting, 2011 IEEE.* IEEE, 2011.

[V] P. Sørensen, B. Andresen, J. Fortmann, K. Johansen and P. Pourbeik,
"Overview, status and outline of the new IEC 61400-27–Electrical simula-
tion models for wind power generation.", *10th International Workshop on
Large-Scale Integration of Wind Power into Power Systems as well as on
Transmission Networks for Offshore Wind Farms*, Aarhus, Denmark, 2011.

[W] R. Hendriks, J. Fortmann, M. Bley, L. Cai and O. Ruhle, "Towards a
Cross-Platform Binary Interface for Time-Domain Simulation Models"
*10th International Workshop on Large-Scale Integration of Wind Power in-
to Power Systems as well as on Transmission Networks for Offshore Wind
Farms*, Aarhus, Denmark 2011.

[X] M. Asmine, J. Brochu, J. Fortmann, R. Gagnon, Y. Kazachkov, C.E.
Langlois, E. Muljadi, J. MacDowell, P. Pourbeik, S. Seman and K. Wiens,
"Model validation for wind turbine generator models." *Power Systems,
IEEE Transactions on* 26.3 (2011): 1769-1782, 2011.

[Y] L. Cai, I. István Erlich and J. Fortmann. "Dynamic Voltage Stability Analysis for Power Systems with Wind Power Plants using Relative Gain Array
(RGA).", *8th Power Plant & Power System Control Symposium (PPPSC)*,
2.-5. September 2012, Toulouse, France, 2012

[Z] F. Shewarega; C. Feltes, F. Koch; J. Fortmann and I. Erlich, "Determination of Dynamic Wind Farm Equivalents Using Heuristic Optimization",

Power and Energy Society General Meeting 2012 IEEE, 22-26 July, 2012, San Diego, CA USA, 2102.

[A4] P. Sørensen, B. Andresen, J. Fortmann and P. Pourbeik , "Progress in IEC 61400-27. Electrical simulation models for wind power generation", *11th International Workshop on Large-Scale Integration of Wind Power into Power Systems as well as on Transmission Networks for Offshore Wind Farms*, Lisbon, Portugal, 2012.

[B3] J. Fortmann and I. Erlich, "Use of a Deadband in Reactive Power Control Requirements for Wind Turbines in European Grid Codes" *11th International Workshop on Large-Scale Integration of Wind Power into Power Systems as well as on Transmission Networks for Offshore Wind Farms*, Lisbon, Portugal, 2012.

[CC] I. Erlich, F. Shewarega, C. Feltes, F. Koch and J. Fortmann, "Offshore Wind Power Generation Technologies", *Proceedings of the IEEE*, Volume: PP , Issue: 99, 2013.

8.3 Patents

[p1] J. Fortmann, H. H. Letas, "Windenergieanlage mit Umrichtersteuerung und Verfahren zum Betrieb", EP000001643609B1

[p2] J. Fortmann, "Verfahren zum Betrieb bzw. Regelung einer Windenergieanlage, sowie Verfahren zur Bereitstellung von Regelleistung mit Windenergieanlagen ",EP000001665494B1

[p3] J. Fortmann, H. H. Letas, "Windenergieanlage mit einem Blindleistungsmodul zur Netzstützung und Verfahren dazu", EP000001668245B1

[p4] J. Fortmann, H. H. Letas, "Windpark mit robuster Blindleistungsregelung und Verfahren zum Betrieb", EP000001802866B1

[p5] J. Fortmann, J. Altemark, J. Zeumer, "Leistungsregelung eines Windparks", EP000001907697B1

[p6] J. Fortmann, H. H. Letas, "Verfahren zum Betreiben einer Windenergieanlage bei Überspannungen im Netz", EP000002135349B1

[p7] J. Fortmann, H. H. Letas, "Windenergieanlage mit doppelt gespeistem Asynchrongenerator und Umrichterregelung", EP000002245717B1

8.4 Patent Applications

[p8] J. Fortmann, H. H. Letas, "Umrichter mit steuerbarem Phasenwinkel", DE102006050077

[p9] J. Fortmann, J. Kretschmann, "Verfahren und Vorrichtung zum Betrieb eines Umrichters, insbesondere für Windenergieanlagen", DE102006053367

[p10] J. Fortmann, H. H. Letas, "Windenergieanlage mit Gegensystemregelung und Betriebsverfahren", DE102006054870

[p11] J. Fortmann, H. H. Letas, "Windpark mit Spannungsregelung der Windenergieanlagen und Betriebsverfahren" DE102007044601

[p12] J. Fortmann, "Windenergieanlagen mit Regelung für Netzfehler und Betriebsverfahren hierfür", DE102007049251

[p13] J. Fortmann, " Windkraftanlage mit Umrichterregelung", DE102008034532

[p14] J. Fortmann, "Verfahren und Stromerzeugungsanlage zum Stabilisieren eines Stromverteilungsnetzes", DE102008062356

[p15] J. Fortmann, "Windenergieanlage und Verfahren zum Betreiben einer Windenergieanlage", DE102010023038

[p16] J. Fortmann, "Windpark und Verfahren zum Betreiben eines Windparks",DE102010056456

[p17] R. Bluhm, J. Fortmann, "Windpark und Verfahren zum Betreiben eines Windparks", DE102010056457

A Appendix

A.1 Space Vectors

The quantities of a three-phase system like voltage, current and flux linkage can be described by a space vector. It consists of two orthogonal and a zero sequence component. The components can be calculated by the Clarke transformation [75] from the tree phase system as

$$
\begin{bmatrix} y_\alpha(t) \\ y_\beta(t) \\ y_0(t) \end{bmatrix} = \frac{2}{3} \begin{bmatrix} 2 & -\dfrac{1}{2} & -\dfrac{1}{2} \\ 0 & \dfrac{\sqrt{3}}{2} & -\dfrac{\sqrt{3}}{2} \\ \dfrac{1}{2} & \dfrac{1}{2} & \dfrac{1}{2} \end{bmatrix} \begin{bmatrix} y_a(t) \\ y_b(t) \\ y_c(t) \end{bmatrix}.
\tag{A.1}
$$

The Clarke transformation is also referred to as $\alpha\beta 0$ transformation, it is usually used for instantaneous values only. If there is no ground connection of the three phase systems, the zero-sequence component is not needed.

In a stationary reference frame without zero-sequence component, the space vector can be written as complex quantity:

$$
\underline{y}^{\angle 0} = y_\alpha + \mathrm{j} y_\beta
\tag{A.2}
$$

The stator oriented coordinate system is then denoted by the exponent $\angle 0$. By using the unit vectors

$$
\underline{a} = -\frac{1}{2} + \mathrm{j}\frac{\sqrt{3}}{2}, \qquad \underline{a}^2 = -\frac{1}{2} - \mathrm{j}\frac{\sqrt{3}}{2},
\tag{A.3}
$$

the space vector can be re-written as

$$
\underline{y}^{\angle 0} = \frac{2}{3}\left(y_a + \underline{a}\, y_b + \underline{a}^2\, y_c \right).
\tag{A.4}
$$

The space vector represents the instantaneous values of the three phase system. Under symmetrical conditions the space vector in stationary coordinates $\underline{y}^{\angle 0}$ rotates at the speed corresponding to the system radian frequency with constant amplitude.

A.1.1   Control Representation

For control purposes, the space vector needs to be converted into a coordinate system rotating with the grid frequency ω. This is denoted by the exponent $\angle \omega$.

$$\underline{y}^{\angle \omega} = \underline{y}^{\angle 0} e^{-j(\omega t + \vartheta_0)} \tag{A.5}$$

If the angle

$$\vartheta = \omega t + \vartheta_0 \tag{A.6}$$

corresponds with the position angle between the α axis and the direct axis of a synchronous machine, the components of the rotating space vectors are denoted as d (direct) and q (quadrature) component.

$$\underline{y}^{\angle \omega} = y_d + jy_q. \tag{A.7}$$

A.1.2   Negative Sequence Representation

During unbalanced conditions, a negative sequence component needs to be considered. It can be represented by a space vector rotating in opposite direction to the positive sequence component. Index 1 is used for positive and index 2 for negative sequence.

By applying the general transformation

$$y = \hat{y}\cos(\omega t + \varphi) = \mathrm{Re}\left\{\hat{y}e^{j\varphi}e^{j\omega t}\right\} = \frac{1}{2}\left(\hat{y}e^{j\varphi}e^{j\omega t} + \hat{y}e^{-j\varphi}e^{-j\omega t}\right)$$
$$= \underline{y} + \underline{y}^* \tag{A.8}$$

a sinusoidal quantity can be represented by a component rotating with grid frequency and a quantity rotating in opposite direction ($-\omega$). Using (A.8), (A.4) can be rewritten as

$$\underline{y}^{\angle 0} = \frac{1}{3}\left[\left(\underline{y}_a + \underline{y}_a^*\right) + \underline{a}\left(\underline{y}_b + \underline{y}_b^*\right) + \underline{a}^2\left(\underline{y}_c + \underline{y}_c^*\right)\right] \tag{A.9}$$

$$\underline{y}^{\angle 0} = \underbrace{\frac{1}{3}\left[\underline{y}_a + \underline{a}\,\underline{y}_b + \underline{a}^2\,\underline{y}_c\right]}_{\underline{y}_1^{\angle 0}} + \underbrace{\frac{1}{3}\left[\underline{y}_a + \underline{a}^2\,\underline{y}_b + \underline{a}\,\underline{y}_c\right]^*}_{\underline{y}_2^{*\angle 0}}$$

$$\underline{y}^{\angle 0} = \underline{y}_1^{\angle 0} + \underline{y}_2^{*\angle 0} = \underline{y}_1^{\angle \omega}e^{j\omega t} + \underline{y}_2^{*\angle -\omega}e^{-j\omega t}$$

with

$$\underline{y}_1^{\angle 0} = y_{1\alpha} + jy_{1\beta} \tag{A.10}$$

$$\underline{y}_2^{\angle 0} = y_{2\alpha} + jy_{2\beta} \tag{A.11}$$

A.1.3   Zero Sequence Components

Zero-sequence (DC) components of the space vector can have a significant impact if positive or negative sequence components alone are to be considered.

$$\underline{y}^{\angle 0} = \underline{y}_1^{\angle 0} + \underline{y}_2^{*\angle 0} + \underline{y}_{DC}^{\angle 0} = \underline{y}_1^{\angle \omega} e^{j\omega t} + \underline{y}_2^{*\angle -\omega} e^{-j\omega t} + \underline{y}_{DC} \qquad (A.12)$$

By multiplying (A.12) with $e^{-j\omega_0 t}$ the positive sequence component has stationary values, while the DC-component rotates with the grid frequency ($-\omega$) and the negative sequence with twice the grid frequency (-2ω). For extracting the positive sequence only, therefore a filter with a lower bandwidth must be used if zero-sequence components are present

$$\underline{y}^{\angle \omega_0} = \underline{y}^{\angle 0} e^{-j\omega_0 t} = \underline{y}_1 + \underline{y}_2^* e^{-j2\omega_0 t} + \underline{y}_{DC} e^{-j\omega_0 t}, \qquad (A.13)$$

The same applies for the negative sequence respectively, if (A.12) is multiplied by $e^{+j\omega_0 t}$. The zero sequence component now rotates with $+\omega$, the positive sequence component with $+2\omega$.

$$\underline{y}^{\angle -\omega_0} = \underline{y}^{\angle 0} e^{j\omega_0 t} = \underline{y}_1 e^{j2\omega_0 t} + \underline{y}_2^* + \underline{y}_{DC} e^{j\omega_0 t}. \qquad (A.14)$$

A.2 Sign Conventions

Generation oriented sign convention is used. Accordingly the counting direction of voltages and currents are defined as shown in Fig. A.1.

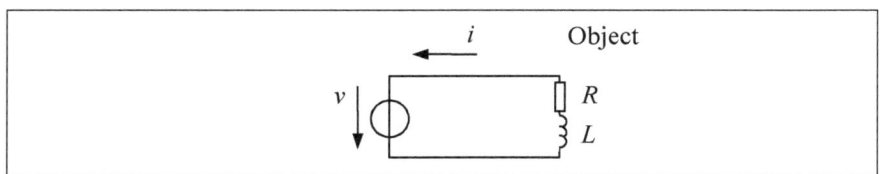

Fig. A.1 Counting direction of voltage and current.

Applying an AC voltage v to the series R-L circuit, shown in Fig. A.1 results in the following voltage equation:

$$v = -R\,i - L\frac{di}{dt}, \qquad (A.15)$$

$$v = -R\,i - \frac{d\psi}{dt}, \qquad (A.16)$$

with

$$\psi = L \cdot i. \tag{A.17}$$

The complex power is defined as

$$\underline{S} = \underline{V}\,\underline{I}^* = \underline{I}\,\underline{I}^*\underline{Z} = P + jQ, \tag{A.18}$$

with

$$P = V_{re}\,I_{re} + V_{im}\,I_{im} = V \cdot I \cos\varphi, \tag{A.19}$$

$$Q = V_{im}\,I_{re} - V_{re}\,I_{im} = V \cdot I \sin\varphi \tag{A.20}$$

and

$$\varphi = \varphi_v - \varphi_i\,. \tag{A.21}$$

and the complex impedance

$$\underline{Z} = \frac{1}{\underline{Y}} = R + jX \tag{A.22}$$

The resulting sign conventions for power are as follows:

Generated active power: positive,
Consumed active power: negative,
Capacitive reactive power: positive,
Inductive reactive power: negative.

The terms "overexcited" and "underexcited" reactive power will also be used to characterize the operating modes of electrical machines. With respect to reactive power in steady state operation, the machine provides capacitive reactive power when it is overexcited and inductive reactive power in underexcited operation

Active and reactive currents are defined as

$$I_P = \frac{P}{V} = \mathrm{Re}\{\underline{I}\} \tag{A.23}$$

$$I_Q = \frac{Q}{V} = -\mathrm{Im}\{\underline{I}\} \tag{A.24}$$

Fig. A.2 clarifies the difference between active and reactive currents on the one side and real and imaginary part of the current on the other.

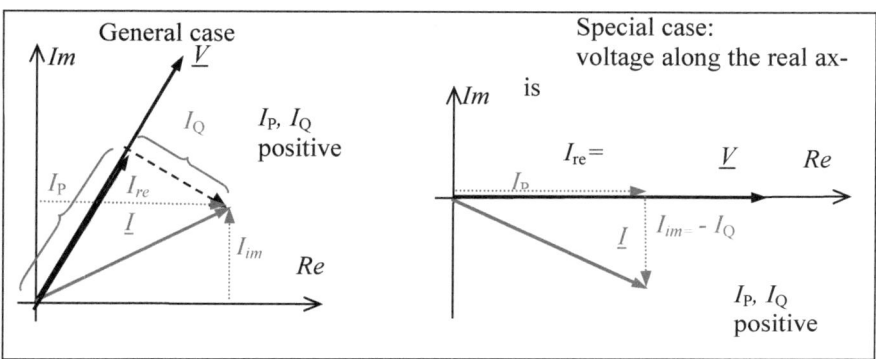

Fig. A.2 Definition of active and reactive current.

A.2.1 Effective Value

Requirements based on the definition of an effective value are still found in technical guidelines. The definition of active power is

$$P_{eff} = \frac{1}{T} \int_{t_0}^{t_0+T} u(t)i(t)\mathrm{d}t \qquad (A.25)$$

A different definition, the half period effective value is sometimes used in regulations [66], [67]. It is defined as

$$P_{eff} = \frac{1}{T/2} \int_{t_0}^{t_0+T/2} u(t)i(t)\mathrm{d}t \qquad (A.26)$$

Effective values based on a full or a half period should not be applied for three phase systems. Especially a calculation of reactive power of three phase systems is not meaningful using effective value calculations of single phases [76].

A.2.2 Symmetrical Components

The definition of active power based on the effective value is given as $P_1 + P_2 + P_3$ as the sum of all phases. The definition of reactive power as given in [77] is the sum of the three phases. But this definition is not sufficient to describe unbalanced conditions [78].

The introduction of symmetrical components (introduced by [79]) defined as

$$
\begin{pmatrix} \underline{I_1} \\ \underline{I_2} \\ \underline{I_0} \end{pmatrix} = \begin{pmatrix} 1 & \underline{a} & \underline{a}^2 \\ 1 & \underline{a}^2 & \underline{a} \\ 1 & 1 & 1 \end{pmatrix} \begin{pmatrix} \underline{I_a} \\ \underline{I_b} \\ \underline{I_c} \end{pmatrix} \tag{A.27}
$$

with $I_{a,b,c}$ as phasors calculated as

$$
\underline{I_a} = i_a(t)e^{j\omega t} \tag{A.28}
$$

allows an improved description of unbalanced conditions.

Methods for calculating the positive and negative sequence for control purposes with a short time delay are described in [48], [80] and [81]. For measurements, an evaluation according to IEC 61400-21 [59] is common.

A.3 Calculation of Symmetrical Components according to IEC 61400-21

The IEC standard 61400-21 [59] describes a method for calculating the positive sequence of a voltage based on instantaneous values. In a first step, the Fourier coefficients for a given frequency can be calculated as

$$
u_{k,\cos} = \frac{2}{T} \int_{t-T}^{t} u_k(t)\cos(2\pi f_1 t)dt \tag{A.29}
$$

$$
u_{k,\sin} = \frac{2}{T} \int_{t-T}^{t} u_k(t)\sin(2\pi f_1 t)dt \tag{A.30}
$$

with $k = a,b,c$

The positive sequence components for voltage and currents result to:

$$
u_{1,\cos} = \frac{1}{6}\left[2u_{a,\cos} - u_{b,\cos} - u_{c,\cos} - \sqrt{3}\left(u_{c,\sin} - u_{b,\sin} \right) \right] \tag{A.31}
$$

$$
u_{1,\sin} = \frac{1}{6}\left[2u_{a,\sin} - u_{b,\sin} - u_{c,\sin} - \sqrt{3}\left(u_{b,\cos} - u_{c,\cos} \right) \right] \tag{A.32}
$$

$$
i_{1,\cos} = \frac{1}{6}\left[2i_{a,\cos} - i_{b,\cos} - i_{c,\cos} - \sqrt{3}\left(i_{c,\sin} - i_{b,\sin} \right) \right] \tag{A.33}
$$

$$i_{1,\sin} = \frac{1}{6}\left[2i_{a,\sin} - i_{b,\sin} - i_{c,\sin} - \sqrt{3}\left(i_{b,\cos} - i_{c,\cos}\right)\right] \qquad (A.34)$$

Positive sequence active and reactive power can then be calculated as

$$p_1 = u_{1,\cos} \cdot i_{1,\cos} + u_{1,\sin} \cdot i_{1,\sin} \qquad (A.35)$$

$$q_1 = u_{1,\cos} \cdot i_{1,\sin} - u_{1,\sin} \cdot i_{1,\cos} \qquad (A.36)$$

A.4 FRT Testing Procedure

According to [53] and [59], FRT – tests should be performed using the measurement setup described in Fig. A.3. A power generation unit (PGU), in this case a wind turbine, is connected to the grid through a testing device. The testing device is designed to reduce the voltage at the PGU to a specified level within a very short period of time.

During normal operation, switch S1 is closed and switch S2 is open. In order to reduce the impact of activating impedance Z2 on the grid, in a first step switch S1 is opened. This connects a serial impedance Z1 in line with the PGU. After the transients have decayed, switch S2 is closed and the short circuit impedance Z2 is connected in parallel to the PGU. This causes a voltage dip at the PGU. After 150 ms .. 2000 ms depending on the tests required, switch S2 is opened again. The voltage at the PGU then recovers. A short time later, switch S1 is closed, and normal operation is resumed.

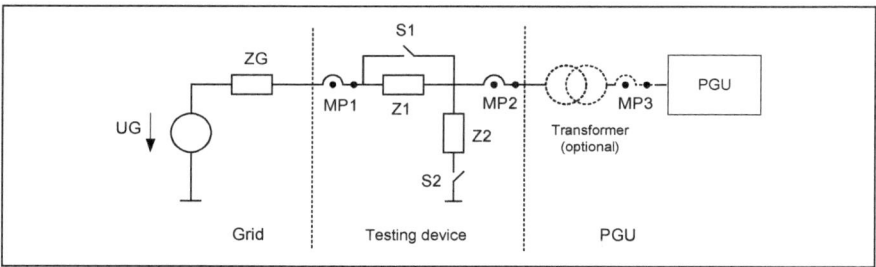

Fig. A.3 Setup to measure the effect of voltage dips on a power generation unit, for example a wind turbine.